First-Order Logic

The Jones and Bartlett Series in Logic and Scientific Method
Gary Jason, *Editor*

Heil, John, Davidson College, *First-Order Logic: A Concise Introduction*

Jason, Gary, San Diego State University, *Introduction to Logic*

The Jones and Bartlett Series in Philosophy
Robert Ginsberg, *General Editor*

Ayer, A. J., *The Origins of Pragmatism: Studies in the Philosophy of Charles Sanders Pierce and William James*

Ayer, A. J., *Metaphysics and Common Sense*, 1994 reprint with introduction by Thomas Magnell, Drew University

Baum, Robert J., University of Florida, Gainesville, *Philosophy and Mathematics*

Ginsberg, Robert, The Pennsylvania State University, Delaware County, *Welcome to Philosophy! A Handbook for Students*

Gorr, Michael, Illinois State University, and Sterling Harwood, San Jose State University, *Crime and Punishment: Philosophic Explorations*

Hanson, Norwood Russell, *Perception and Discovery: An Introduction to Scientific Inquiry*

Heil, John, Davidson College, *First -Order Logic: A Concise Introduction*

Jason, Gary, San Diego State University, *Introduction to Logic*

Pauling, Linus, and Ikeda, Daisaku, *A Lifelong Quest for Peace: A Dialogue* Translated and Edited by Richard L. Gage

Pojman, Louis P., The University of Mississippi, *Environmental Ethics: Readings in Theory and Application*

Pojman, Louis P., The University of Mississippi, *Life and Death: Grappling with the Moral Dilemmas of Our Time*

Pojman, Louis P., The University of Mississippi, *Life and Death: A Reader in Moral Problems*

Pojman, Louis P., The University of Mississippi, and Francis Beckwith, University of Nevada Las Vegas, *The Abortion Controversy: A Reader*

Rolston III, Holmes, Colorado State University, *Biology, Ethics, and the Origin of Life*

Veatch, Robert M., The Kennedy Institute of Ethics, Georgetwon University, *Cross-Cultural Perspectives in Medical Ethics: Readings*

Veatch, Robert M., The Kennedy Institute of Ethics, Georgetown University, *Medical Ethics*

First-Order Logic

A Concise Introduction

John Heil

Davidson College, North Carolina

Jones and Bartlett Publishers
Boston   *London*

Editorial, Sales, and Customer Service Offices

Jones and Bartlett Publishers
One Exeter Plaza
Boston, MA 02116
1-617-859-3900
1-800-832-0034

Jones and Bartlett Publishers International
P. O. Box 1498
London W6 7RS
England

Library of Congress Cataloging-in-Publication Data
Heil, John
    First-order logic : a concise introduction / John Heil.
        p. cm.
    Includes index.
    ISBN 0-86720-957-7
    1. First-order logic. I. Title.
    BC128.H45   1994
    160--dc20   94-3113
                                                          CIP

Acquisitions Editors: Arthur C. Bartlett and Nancy E. Bartlett
Manufacturing Buyer: Dana L. Cerrito
Cover Designer: Marshall Henrichs
Text Printer: Edwards Bros., Inc.
Cover Printer: New England Book Components, Inc.

Printed in the United States of America
98 97 96 95 94   10 9 8 7 6 5 4 3 2 1

For Katherine and Johnny

Contents

CHAPTER FOUR *The Language L_p*

CHAPTER FIVE *Derivations in L_p*

Preface

Why another logic textbook? Why indeed. The market is flooded with textbooks, each of which fills, or purports to fill, a particular niche. Oddly, in spite of—or perhaps because of—the availability of scores of textbooks, many teachers of logic spurn commercial texts and teach from notes and handouts. This suggests that although there are many logic textbooks, there are not many *good* logic textbooks.

Logic texts fall into two categories. Some, like S. C. Kleene's *Mathematical Logic* (New York: John Wiley and Sons, 1967) and Benson Mates's *Elementary Logic* (New York: Oxford University Press, 1972), emphasize logic as a distinctive subject matter to be explicated by articulating, as elegantly as possible, the theory on which the subject matter rests. Others, too numerous to mention, focus on *applications* of logic, treating logic as a skill to be mastered, refined, and applied to arguments advanced by politicians, editorial writers, and talk show hosts. A few authors offer a middle ground, notably E. J. Lemmon in *Beginning Logic* (originally published in 1965, reissued by Hackett in 1978), and Paul Teller in his two-volume *A Modern Formal Logic Primer* (Englewood Cliffs: Prentice-Hall, 1989). Lemmon and Teller embed discussions of theory within a context that encourages the development of logical skills.

In this volume, I have elected to take this middle way. I focus on the construction of translations and derivations, but locate these within a broader theoretical framework. The book assumes no prior contact with, or enthusiasm for, formal logic. My aim has been to introduce the elements of first-order logic gradually, in small steps, as clearly as possible. I have tried to write in a way that is, I hope, congenial to students (and instructors) who ordinarily feel uncomfortable in symbolic domains. My approach to logic is not that of a card-carrying logician. This, I think, gives me something of an advantage in understanding what nonlogicians and symbolphobes find difficult or unintuitive. As a result, I spend more time explaining fundamental notions than other authors do. In my view, this pays dividends in the long run.

The volume covers elementary first-order logic with identity. I have not attempted to offer proofs for the soundness and completeness of the systems introduced. I have, however, offered sketches of what such proofs involve. These are included, with materials on the syntax and semantics of the systems, in sections on metalogic at the end of chapters two through five. These sections can be skipped without loss of continuity. They are offered as springboards for more elaborate classroom discussions. For my own part, I think it important to include a dose of metalogic in an introductory course. Metalogic brings order to materials that are apt to seem arbitrary and ad hoc otherwise. Less obviously, an examination of the syntax, semantics, and metatheory of a formal system tells us something about ourselves. In mastering a formal system we come to terms with a domain that can be given a precise and elegant description. Any account of our psychology, then, must allow for our ability to understand and deploy systems with these formal properties.

The book began life in the summer of 1972. I received support from the National Endowment for the Humanities to write a text that would combine logic with work in linguistic theory. My claim was that this was a case in which learning two things together was easier, more efficient, and more illuminating than learning either separately. The project culminated in a photocopied text inflicted on successive generations of students. In the ensuing years, linguists progressed from transformational grammar, through generative semantics, and on to government binding theory. My interest in the linguistic theory waned, as did my enthusiasm for combining logic and linguistics in a single package. Meanwhile, the manuscript went through a series of photocopied incarnations. Although each version differed from its predecessor only a little, the cumulative effect has been massive.

Along the way, I received help from many people. The original project was inspired by John Corcoran, many of whose ideas I filched shamelessly. I owe a large debt, as well, to David Zaret who kindly read and offered advice on sections devoted to metalogic. Joseph DeMarco, Robert Ginsburg, Kathleen Smith, and two unnamed referees read an earlier version of the entire text and furnished countless criticisms and suggestions, all of which have improved the finished product. Many of my students provided suggestions, corrections, and advice. I am particularly grateful to Susan Peppers for reading and commenting on a large portion of the final draft. Barbara Hannan and Susan Moore caught numerous mistakes and infelicities in early versions put to use at Randolph-Macon Woman's College.

Angela Curran and Robert Stubbings provided indispensable help in ironing out technical difficulties at different stages of the project. My Dean, Robert C. Williams, generously provided a computer, and Gary Jason a study guide designed to accompany this book. My greatest debt is to Harrison Hagan Heil, who brought me to appreciate the importance of saying clearly what can be said clearly.

Berkeley, California   *January, 1994*

1

Introduction

Why study logic? The question is bound to be asked by anyone embarking on a course in elementary logic. The best answer is that there is no simple, uniform answer. Why study Economics? or History? or Biology? There are various reasons ranging from the mundane—"I need to satisfy a distribution requirement," "I need an easy *B*"—to the sublime—"It's fascinating, important, and I can't imagine life without it."

Logic strikes some of us as intrinsically fascinating and important. With luck, a measure of this fascination and importance will emerge in the chapters to follow. Even if the study of logic had little or no intrinsic significance, however, there would still be endless reasons for pursuing it.

Consider, first, the remarkable fact that human beings possess a capacity to learn and deploy languages. Aristotle (384–322 B.C.) characterized human beings as rational animals. The brand of rationality that sets us off from other creatures is linked in important ways to our linguistic endowment. Language is required for the expression—and perhaps even for the entertaining—of thoughts that make possible the coordinated social activities that give meaning to our lives. Faithful dog Spot, we say, thinks his master is at the door. But how plausible is it to imagine Spot thinking—*now*—that his master will be at the door two days hence? A mute, languageless creature like Spot evidently lacks the means to express such a thought. It does not follow immediately that Spot could not, even so, *entertain* the thought. That seems right. But ask yourself what reason we could have for ascribing such a thought to Spot. Arguably, thoughts about a subject matter outside a thinker's immediate environment are, in one way or another, dependent on a linguistic background. Thoughts about spatially or temporally remote states of affairs, thoughts about abstract entities (like numbers or classes), and thoughts about nonexistent things (unicorns, mermaids, the Easter Bunny) seem, on the face of it, to require a linguistic medium.

1

As we shall see, the study of logic provides insights into the structure of natural languages—the languages we speak every day. That language might or might not be English. For purposes of logic, it does not matter. The logical forms we shall study have much in common with abstract logical forms underlying all natural languages. The study of logic, then, can reveal much about the character of language. Given the central place of language in human affairs, the study of logic can help illuminate human psychology.

Talk and Thought

Suppose you observe Spot nosing about quietly on Monday and Tuesday, then, on Wednesday, pacing expectantly in the vicinity of the front door. Would this entitle you to conclude that, on Monday and Tuesday, Spot entertained the thought that his master would return on Wednesday? Or does Spot's behavior suggest, rather, that *on Wednesday* Spot thinks his master's arrival is imminent? What reasons could you have, here and now, for taking Spot to be entertaining a *tensed thought*, a thought about the future, one we would express by means of the English sentence, "Master will return the day after tomorrow"?

The history of logic begins with Aristotle's classification of argument forms. Arguments are collections of sentences, one of which, the *conclusion*, is supported by one or more *premises*. Arguments exhibit repeatable *patterns*. Some of these patterns represent valid reasoning—their premises imply their conclusions—and some do not. (We shall see later how to define precisely the notion of implication. For the moment, we can suppose that premises imply a conclusion when they provide support for that conclusion: if we accept the premises, we have reason to accept the conclusion.)

Aristotle focused on simple *syllogistic* patterns of reasoning, those involving a pair of premises and a conclusion. He recognized that many of the arguments used by philosophers, politicians, scientists, and ordinary people could be understood as exhibiting combinations of these patterns. He reasoned that if an argument is valid, if its premises provide grounds for its conclusion, then any argument with the same pattern must be valid as well. This nicely illustrates the *formal* character of logic. Logic provides a way of studying and classifying repeatable forms or patterns of reasoning.

We shall discover that formal logic provides a powerful technique for assessing the validity and invalidity of arguments. This is bound to prove useful when we turn to the examination of arguments in ordinary life, the stuff of editorial pages and debates over public policy. It proves useful, as well, in the evaluation of theories about the world and our place in it. Logic provides a framework that clearly exhibits the structure of lines of reasoning. To the extent that we can transform everyday reasoning into a standard logical form, we *discipline* that reasoning. In so doing, we may find that cherished beliefs are based on specious inferences—or that they are better supported than we had imagined.

Syllogistic Patterns of Reasoning

These two arguments exhibit similar *forms* or *patterns* of inference:

All whales are mammals	All philosophers are logicians
All mammals are mortal	All logicians are clever
All whales are mortal	All philosophers are clever

The sentences above the horizontal line, premises, are taken to support the sentence below the line, the conclusion. These arguments are *valid*: their premises imply their conclusions. Must valid arguments have true premises? True conclusions?

Some philosophers have thought that elementary formal logic of the sort we shall be studying provides a *canonical notation* for the pursuit of knowledge. The idea is simple and elegant: anything we can intelligibly think or say about the world must be expressible in this favored idiom. Whatever is not so expressible would not be a candidate for truth *or* falsehood. If that is right, there is much to be learned about the structure of our conception of reality by studying logic. Even if the idea of a canonical language is off-base, formal logic provides insight into one central aspect of the way we think about the world and our place in it.

Whatever its standing in the broader scheme of things, logic has deep ties to mathematics and computer science. Anyone bent on pursuing serious work in either of these areas stands to benefit from the study of logic. Other disciplines, too, are connected to formal logic in fundamental ways.

Quantum physics is sometimes held to mandate a nonstandard logic for the description of quantum events. The claim can be sensibly evaluated only against a background of a standard logic of the sort we shall take up in the chapters to follow.

I have thus far omitted mention of one reason widely cited for taking up the study of logic. By working at logic, we might expect to enhance our reasoning skills, thereby improving our performance on cognitive tasks generally. I have not emphasized this supposed benefit, however, because I am skeptical that there is much to it. The empirical evidence casts doubts on the notion that training in logic leads to improvement in ordinary reasoning tasks of the sort we encounter outside the classroom. Formal logic, like most other learned disciplines, resists "transference" across problem domains. People trained in statistics, for instance, fare no better than the rest of us at "real world" tasks requiring reasoning under uncertainty. Similarly, students with *A*'s in logic may continue to make errors in everyday reasoning. The fault lies not with logic, but with our tendency to compartmentalize what we learn.

Does logic, then, afford *nothing* in the way of overall improvement in reasoning? That, too, is doubtful. We should scale down our expectations and recognize that improvement in reasoning prowess, if any, is likely to be incremental, rather than dramatic. In studying logic, we are more apt to become sensitive to species of *bad* reasoning; and when problems are framed in ways that bring out their connections to the domain of formal logic, our training can serve us well.

1.01 MAKING THE MOST OF THIS BOOK

I have noted a variety of *extrinsic* reasons for taking up the study of logic. You may, of course, find the subject challenging and satisfying in itself. When this is so, logic is *exciting*. My aim is to present materials in a way that will draw you in and enable you to experience that excitement.

The success of the enterprise will require cooperation on your part. Mastery of formal logic comprises mastery of a range of *skills*. Logic encompasses a subject matter, but it includes as well techniques for the *perspicuous representation* of sentences in natural languages—that is, the translation of ordinary sentences into a formal notation that reveals their fundamental logical structure—and for the construction of derivations. Both sentence representation and derivation construction require *practice*. Like practice at the piano, what is at first difficult and alien can evolve into something obvious and familiar.

The book is sprinkled with exercises designed to encourage practice on skills required for devising translations and constructing derivations. These exercises—and their components—vary in difficulty. Most are simple, a few are more demanding. In every case, their value hinges on your working through them systematically. You will discover that there is a vast difference between, on the one hand, *recognizing* or *understanding* the translation of a sentence into a formal idiom or the construction of a derivation, and, on the other hand, *translating* the sentence or *constructing* the derivation yourself. Translation and derivation construction require skills and insight gained from practice, not simply the exercise of recognitional abilities.

The difference here is strictly analogous to that associated with any skillful activity. I can *hear* that a piece is played correctly on the piano or *see* that a tennis serve is properly executed. It scarcely follows that *I* could play the piece or properly serve a tennis ball. To do so, I should need to practice, to repeat movements until they came smoothly and naturally. In mastering techniques in formal logic, you must be prepared to repeat perceptual and cognitive maneuvers until they become routine. To that end, you will need to rework—and to re-rework—sentence translations and derivations that strike you as troublesome. In so doing, you may imagine that you are learning nothing new. In one sense that is so. In practicing a piece on the piano, nothing new is learned; instead, a skill is enhanced. So it is with translations and derivations. Repetition improves and fine-tunes the pertinent skills.

Formal techniques used to devise translations and derivations include *perceptual* as well as *cognitive* ingredients. That may surprise you. You may imagine that mastery of formal logic is mostly a cognitive or intellectual achievement. Some of us are said, like Spock, to "think logically," although most of us think in other ways. Perhaps logicians and mathematicians are predominantly "left-brained," and the rest of us "right-brained." This picture is, I am convinced, both wrong and misleading.

The steps required in the construction of derivations are, as we shall see, invariably trivial. They are moves we master in childhood. We know, for instance, that if all birds fly, and robins are birds, then robins fly. Success in logic requires little more in the way of inferential sophistication. The challenge is, instead, perceptual. You must learn to *see* patterns within arrays of symbols. Once you can do this, it will be a simple matter to apply inferential strategies required for the construction of derivations. Seeing patterns among symbols requires perceptual skills of the sort we deploy when we identify species of bird or kinds of tree. Picture Wayne, embarking on a leaf-gathering expedition for his eighth-grade biology class, armed with the *Audubon Field Guide to North American Trees*. At first Wayne finds it difficult or impossible to identify trees by comparing what he sees to pictures and descriptions in the *Guide*. The *Guide* displays a picture of an elm leaf. Does the leaf in front of Wayne now match that picture? Well, it does in certain respects. It has a notch, however, where the pictured leaf does not. Is the notch damage done by an insect, or is it natural? If it is natural, is it uncharacteristic of elm leaves? Or do elm leaves, perhaps, vary in respect to their notches?

With practice, Wayne will eventually acquire the knack of recognizing elm leaves. His doing so depends on his having come to appreciate what is and what is not relevant to a leaf's being an elm leaf. The knack is not one he could easily put into words. (Imagine trying to explain to someone over the telephone how to identify species of leaf.) In this regard, the perceptual skill Wayne has acquired resembles the motor skills we master when we learn walk, tie a bow, ride a bicycle, or ski. We learn such things by *doing* them—badly, at first, then, with practice, better, until, eventually, we perform them automatically. Early on, we concentrate: *to turn, keep your weight on the downhill ski.* Later, we simply ski. The motor routines involved have been programmed to run without conscious direction.

So it is with the construction of translations and derivations. What is difficult at first becomes, with practice, perceptually a snap. Practice is essential and unavoidable. Some have a headstart on this. We may be inclined to regard such people as smarter or "more logical" than others. The abilities in question, I have suggested, are less intellectual than perceptual, less a matter of cogitation than of skillfully seeing patterns. If pattern recognition is "right-brained," the skills required for success in logic are, perhaps surprisingly, "right-brained" skills.

Practice and repetition enable us to *automatize* conscious processes. This is a vital component in our mastery of any subject matter. The opposite process—that is, bringing to conscious awareness what we now do unself-consciously—is equally vital. In speaking and understanding a language, we make use of a variety of *syntactic* and *semantic* skills. Roughly, *syntax* designates the formal grammatical structure of sentences; *semantics* captures their meanings. When we set out to translate a particular English sentence into a formal idiom, we must first make explicit to ourselves the semantics of that sentence. Only then can we be sure we have found a plausible formal counterpart. Our semantic knowledge, however, like the knowledge we have of walking, shoe-tying, and skiing, is largely *implicit*. We have it, but it is not easily articulable. The study of logic encourages us to make explicit what we should otherwise make use of without much awareness of what we are doing. In the end, we stand to learn much about ourselves.

In the pages that follow, we shall take up two formal systems, two *languages*. I dub these L_s and L_p. L_s is a simple *sentential* logic, a system whose elements consist of "atomic" sentences and sentential connectives that allow for the construction of complex "molecular" sentences. L_p is a *predicate* logic. Its elements include subatomic parts of sentences. Because sentential logics are simpler than predicate logics—indeed, L_p incorporates L_s as a subpart—we shall begin with L_s, and then move to a consideration of L_p. Throughout it all, you should bear in mind the importance of perceptual skills essential to the mastery of material under discussion. You should bear in mind, as well, the importance of making clear to yourself what you, in one sense, already know about English. Thus prepared, you will be in a position to discover both the austere beauty exhibited by formal systems and the limitations of these systems as vehicles of thought and meaning.

2

The Language $\mathcal{L}_s$

2.00 A FORMAL LANGUAGE

This chapter introduces a simple formal language, $\mathcal{L}_s$. Although $\mathcal{L}_s$ is simple in comparison to natural languages like English, German, or Urdu, it exhibits a variety of interesting logical properties. We focus first on its syntax and semantics, then, in the next chapter, on its proof structure. Throughout the discussion, features of $\mathcal{L}_s$ will be compared with those found in English. We may learn something about the logical character of a familiar natural language in the course of examining a simplified artifact.

Like every language, $\mathcal{L}_s$ incorporates an elementary *vocabulary* from which sentences are constructed. Sentences can be as complex as we choose to make them, so long as their complexity remains finite. There is no longest sentence in $\mathcal{L}_s$, but its sentences (or those of English, for that matter) cannot be *infinitely* long. Sentences of $\mathcal{L}_s$, or of any natural language, are *finite* strings of elements: sentences are *finitely constructible*. The significance of these points will become clear in the course of our discussion of the syntax of $\mathcal{L}_s$ (§ 2.20).

Admittedly, ordinary human beings would be baffled by excessively complex sentences or those exceeding a certain length. Such matters fall within the province of the psychologist, however; they do not affect our characterization of a language. Mathematicians are interested in defining and exploring the properties of mathematical systems, although the *uses* of such systems and the cognitive difficulties that human beings might experience in dealing with them are not themselves mathematical issues. The same holds for a logician aiming to characterize a language like $\mathcal{L}_s$, or the linguist interested in characterizing English. In each case the goal is a description of an "abstract entity," a language. Properties of such entities are not to be mistaken for properties of those who employ them.

2.01 SENTENTIAL CONSTANTS AND SENTENTIAL VARIABLES

English sentences are made up of nonempty but finite strings of words arranged in a particular order. Every sentence is made up of at least one word, and no sentence includes an infinite number of words. The building blocks of L_s sentences, in contrast, are not words, but elements that themselves function as sentences. These elements, the *sentential constants*, are represented by means of familiar uppercase letters, A through Z. Sentential constants behave in L_s in roughly the same manner that simple declaritive sentences behave in English. They can occur individually or together with other sentential constants within complex sentences. The L_s sentence, S, then, resembles the English sentence "Socrates is wise." Each sentence can be used to express a simple proposition. Like its English counterpart, the L_s sentence can be combined with other simple sentences (in ways to be discussed) to produce sentences expressing complex propositions.

Elements of L_s, like those in natural languages, do not come with built-in meanings. *We* decide what a given element is to mean, how it is to be interpreted. In the case of natural languages, this decision may be conscious and deliberate, as when a scientist invents a term to designate a newly discovered particle, or it may be the result of a tacit social agreement shrouded in history. In learning a language we enter into an implicit agreement with others sharing the language, an agreement that enables us to use words in a way that reliably communicates what we intend to communicate.

Formal languages like L_s serve very different functions. We use L_s to make clear the logical structure of sentences and logical relations among sentences in natural languages. As a result, it is possible to restrict the elements of L_s dramatically. Any sentence of L_s can be constructed from a small number of simple sentences, because we can elect to interpret these simple sentences differently on different occasions. (It is possible to supplement the elementary vocabulary of L_s by introducing subscripts: A_1, A_2,... A_{519}.... This is a complication that need not concern us here; see § 4.14.) On one occasion, then, we may use the L_s expression, S, to mean

Socrates is wise.

On another occasion, we might use the very same symbol, S, to mean

Socrates is happy.

We can make life easier for one another by selecting letters that bear some relation to corresponding English sentences. In the examples above *S* was used to express propositions expressed, respectively, by the English sentences "Socrates is wise" and "Socrates is happy." We might just as well have used *W* and *H*. The choice of letters is constrained, not by logic, but by common sense: we translate sentences in a way that will be easy for us to remember and for others to decipher. We cannot, however, use the same letter in a single complex sentence to express two *distinct* propositions.

Sentences and Propositions

Philosophers commonly speak of sentences as *expressing propositions*. Sentences vary from language to language. Propositions, in contrast, are taken to possess a kind of language-independent meaning.

Distinct sentences can be used to express the same proposition: the English sentence "Snow is white" expresses the same proposition expressed by the French sentence "La neige est blanche." The same sentence can, on different occasions, be used to express different propositions: the English sentence "They are flying planes" could be used to express two different propositions. What are they?

Although it is undeniably convenient to appeal to propositions in discussing language—we can say, for instance, that a sentence in one language translates a sentence in another language if both are used to express the same proposition—we should bear in mind that there is little agreement among philosophers as to what propositions are, or even whether such things exist.

Suppose we set out to represent in $\mathcal{L}_s$ something equivalent to the English sentence

Socrates is wise and Socrates is happy.

We cannot represent *both* simple constituent sentences—the sentence "Socrates is wise" and the sentence "Socrates is happy"—by means of an *S*. We must instead use distinct letters, *W* and *H*, for instance.

Occasionally it will be necessary to talk in a general way about properties of $\mathcal{L}_s$ sentences. We might, for instance, want to discuss sentence *sorts*, or sentential structures generally, without restricting ourselves to particular sentences. Faced with a similar need, mathematicians resort to *variables*. In explaining the operation of multiplication, for instance, I could set out the following characterization:

> Where x and y are integers, their product is
> equal to x added to itself y times.

Here x and y function as variables ranging over arbitrary numbers. In discussing $\mathcal{L}_s$, we make use of a set of *sentential variables* that range over arbitrary $\mathcal{L}_s$ sentences. Sentential variables consist of the *lowercase* letters, p through u. The sentential variable, p, then, might be used to stand for any sentence of $\mathcal{L}_s$ we choose. The introduction of such variables is more than a convenience. Without them, we would be unable to describe $\mathcal{L}_s$ and its properties in a suitably general way. Variables must not be mistaken for sentences, however. Just as x and y are not numbers, so p and q are not sentences of $\mathcal{L}_s$. In writing sentences in $\mathcal{L}_s$, then, we use *only* uppercase letters.

Exercises 2.00

Provide $\mathcal{L}_s$ translations of the English sentences below.

1. Socrates is brave.

2. Socrates loves Xantippi.

3. Xantippi loves Socrates.

4. The quick brown fox jumped over the lazy dog.

5. Every good boy does fine.

2.02 TRUTH FUNCTIONAL CONNECTIVES

The richness of a natural language like English arises from its capacity to yield an unlimited supply of sentences from a finite vocabulary. Mastering a language involves mastering techniques for producing and understanding sentences belonging to that infinite stock. An elementary technique for generating sentences consists of combining simple sentences into compounds. We take the sentences "Socrates is wise" and "Socrates is happy" and put them together to form the compound sentence

Socrates is wise and Socrates is happy.

In the course of assembling simple sentences to produce compounds, we often modify the originals in a way that disguises their structure. Thus, although we can combine the sentences "Socrates is wise" and "Socrates is happy" to yield the sentence above, we are more likely produce something like

Socrates, who is wise, is happy

or perhaps

Socrates is wise and happy.

In the first example, the sentence "Socrates is wise" is converted to a relative clause, "who is wise," and *embedded* inside the sentence "Socrates is happy." In the second example, elements in one sentence that are repeated in the other sentence have been dropped. Such processes are common in natural languages. We shall examine them in more detail presently.

In using English, we combine simple sentences to form larger, more informative sentences. This holds for L_s as well. There are, however, important differences. First, as the examples above illustrate, English sentences constructed from simpler English sentences typically contain *transformations* of the original simple sentences. A sentence, combined with another, might be converted into a clause, or have its repeated elements dropped. In L_s, simple sentences maintain their identity. This is one reason a formal language like L_s can be taken to reveal logical structure hidden or disguised in sentences used in natural languages. A second difference between L_s and English is that the mechanism allowing for sentential combination in L_s is more restricted than the combinatory mechanisms typical of natural languages. In this chapter, five *truth functional sentential* connectives (sometimes called *logical connectives*, *logical constants*, or *logical operators*) will be introduced. These serve to link simple $\mathcal{L}_s$ sentences together. Truth functional connectives, unlike their natural language counterparts, have no effect on the structure of the sentences they bind together. Complex $\mathcal{L}_s$ sentences will be obvious compounds of simple $\mathcal{L}_s$ sentences.

As a reflection of this feature of $\mathcal{L}_s$, simple $\mathcal{L}_s$ sentences are called *atomic sentences*, and distinguished from compound *molecular sentences*. Molecules in nature are made up of atoms as parts. In making up a molecule, atoms retain their identity. Similarly, molecular sentences in $\mathcal{L}_s$ are made up of elements that keep their sentential identity. We have seen that natural languages are, in this respect, very different. In constructing a complex sentence, we typically transform the structure of its simple constituents, often beyond recognition. This feature of English, a feature we all exploit constantly and unreflectively, can lead to difficulties when we set out to translate from English into $\mathcal{L}_s$. In constructing a translation, in finding an $\mathcal{L}_s$ sentence that "matches" some English sentence, we must recover information no longer obviously present in the sentences we are given.

2.03 NEGATION: ¬

The ¬ symbol will be used to represent negation in $\mathcal{L}_s$. By placing this symbol to the left of an $\mathcal{L}_s$ sentence, we negate it in much the same way we might negate an English sentence by placing to its left the phrase "it's not the case that...." Suppose we designate the $\mathcal{L}_s$ sentence, W, to mean

Socrates is wise.

The sentence ¬W would mean

It's not the case that Socrates is wise,

or, more colloquially,

Socrates isn't wise.

Similarly, suppose we use H to mean

Socrates is happy.

In that case, its negation, ¬H would mean

It's not the case that Socrates is happy,

that is,

Socrates isn't happy.

Negation is the first of five *truth functional sentential connectives* we shall discuss. By comparing its operation to the phrase "it's not the case that…" in English, we obtain an informal grasp of its significance. We can, however, characterize its significance more precisely by means of a *truth table*.

╭──────── Truth Functional Connectives ────────╮

$\mathcal{L}_s$ makes use of five truth functional connectives:

¬ negation (it's not the case that…)

∧ conjunction (…and…)

∨ disjunction (either…or…)

⊃ conditional (if…then…; …only if…)

≡ biconditional (…if and only if…)

╰──╯

Sentences in $\mathcal{L}_s$ take on one of two values: true (*T*) or false (*F*). We can define truth functional connectives by providing tables—*truth tables*—that indicate the effect of a connective on the truth values of sentences in which they occur. The table below characterizes the negation connective:

p	$\neg p$
T	F
F	T

Notice that this table makes use of a sentential variable, p, rather than some particular sentence. This endows the definition with a level of generality it would otherwise lack. The truth table tells us that, given *any* $\mathcal{L}_s$ sentence, p, if p is true, then $\neg p$ is false; and if p is false, then $\neg p$ is true. Negation, then, *reverses the truth value* of $\mathcal{L}_s$ sentences. In this regard, negation in $\mathcal{L}_s$ resembles negation in English. If the sentence "Socrates is wise" is true, then "It's not the case that Socrates is wise" (or "Socrates isn't wise") is false; and if the original sentence is false, its negation is true.

$\mathcal{L}_s$ is said to be a *truth functional language*: the truth value of every sentence in $\mathcal{L}_s$ is a function of the truth values of its constituent sentences. In practice, this means that given any $\mathcal{L}_s$ sentence, p, we can precisely determine its truth value—say whether it is true or false—if we know (1) the truth values of its constituent sentences, and (2) the definitions of the truth functional connectives. If we know that the sentence A is false, then we know (given the definition above) that $\neg A$ is true, and so on for every sentence in $\mathcal{L}_s$.

Truth Functions

$\mathcal{L}_s$ is a *truth functional language*, a language in which the truth value of every sentence is a function of—is completely fixed by—the truth value of its constituent sentences. If you know the truth values of the simple sentences, then you can easily determine the truth values of any complex sentence in which those simple sentences figure.

The connectives in $\mathcal{L}_s$ ($\neg$, $\wedge$, $\vee$, $\supset$, $\equiv$) are truth functional connectives. This means that they are defined by reference to the effects they have on the truth values of sentences in which they occur.

We can describe $\neg p$ or $p \wedge q$ as functions—truth functions—just as we can describe x^2 as a mathematical function.

2.04 CONJUNCTION: $\wedge$

A second truth functional connective represents the operation of *conjunction*. It is symbolized in $\mathcal{L}_s$ by the nameless symbol, $\wedge$. Conjunction in $\mathcal{L}_s$ mirrors conjunction in English. If we place a $\wedge$ between two sentences, we obtain a new compound sentence. Thus from W and H, we may obtain the conjunction $W \wedge H$. In English, we arrive at the conjunction

Socrates is wise and Socrates is happy

by using "and" to conjoin the sentences "Socrates is wise" and "Socrates is happy." We are free to combine negated with non-negated sentences or with other negated sentences to form more complex conjunctions

$$\neg W \wedge H$$
$$W \wedge \neg H$$
$$\neg W \wedge \neg H$$

All of these are perfectly acceptable sentences, as are

$$(\neg W \wedge \neg H) \wedge A$$
$$(W \wedge \neg H) \wedge (A \wedge B)$$

sentences built up from more than two atomic components. In general, we can assemble conjunctions of any finite length. In each case the truth value of the resulting sentence will be a function of the truth values of its constituents.

A truth table characterization of the $\wedge$ is set out below:

p	q	$p \wedge q$
T	T	T
T	F	F
F	T	F
F	F	F

This truth table is more complicated than the truth table used to characterize negation. There, we needed to specify the action of negation on single sentences: the negation of any sentence results in a reversal of the truth value of that sentence—from true to false, or from false to true. Since we needed only a *single* sentential variable, every possible truth value of sentences over which that variable ranged could be specified by two rows in the table: a sentence to which a negation sign is appended may be true or false. Because conjunction is used to conjoin *pairs* of sentences, its truth table requires additional rows to allow for the specification of every possible truth value combination of the conjoined sentences. Sentences flanking the $\wedge$ are called *conjuncts*. Every use of the $\wedge$ involves a pair of conjuncts, so there are *four* possible combinations of truth values to be considered: (1) both conjuncts might be true; (2) the first might be true, the second false; (3) the first might be false, the second true; or (4) both conjuncts might be false.

Functions

Truth tables resemble function tables, which allow us to depict in a tabular way the action of a particular function. The squaring function, for example, might be pictured by means of the following table:

x	x^2
0	0
1	1
2	4
3	9
4	16
5	25
⋮	⋮

The values appearing on the left side of the table represent the *domain* of the function; those to the right represent its *range*. Functions provide mappings from a domain to a range: they associate elements in the one with elements in the other. In the table above, elements in the set of positive integers are associated with elements of that same set. Truth tables map or associate truth values with truth values.

The truth table on p. 16 exhibits the truth value of the resulting conjunction given each of these combinations of values for its conjuncts. A conjunction is true only when *both* of its conjuncts are true (the situation illustrated by the table's first row). In every other case, the resulting conjunction is false—as the remaining rows indicate. This feature of conjunction in $\mathcal{L}_s$ resembles conjunction in English. In general, when English sentences are joined by "and," the resulting compound sentence is true if *both* of its constituent sentences are true, false otherwise. The sentence

Socrates is wise and happy

is true when and only when the sentence "Socrates is wise" and the sentence "Socrates is happy" are both true. (Recall that the English sentence above is a stylistic variant of the sentence "Socrates is wise and Socrates is happy.")

2.05 SENTENTIAL PUNCTUATION

Before delving further into the mysteries of $\mathcal{L}_s$, let us reflect briefly on a problem of notational ambiguity. Suppose you are asked to find the value of the arithmetical expression

$$2 + 3 \times 5$$

In the absence of additional information, the expression is *ambiguous*, that is, it might mean

the sum of 2 and 3, times 5 $(= 25)$

or it might mean

2 added to the product of 3 and 5 $(= 17)$.

The difference in the values of these readings of the original expression illustrates the reason mathematicians can ill afford ambiguity. To avoid ambiguity, we adopt various notational conventions. We might decide, for instance, always to perform operations in a left-to-right sequence. Were we to follow this convention in the example above, we should interpret the expression in the first way. Alternatively, we might adopt a system of punctuation that made use of right and left parentheses: "(" and ")". That is, we could insert parentheses so as to force one or another reading

$$(2 + 3) \times 5$$

or

$$2 + (3 \times 5)$$

The rule, were we to take the trouble to formulate it, is that expressions occurring inside matching parentheses are to be replaced by the values of which they are functions. Thus, $(2 + 3)$ is to be replaced by 5, and (3×5) by 15.

In the case of $\mathcal{L}_s$, we shall adopt a similar technique. Consider the sentence

$$\neg P \wedge Q$$

Is this expression to be read as the negation of P, $\neg P$, conjoined to Q? Or ought we read it as the negation of the *conjunction*, $P \wedge Q$? Let us introduce a convention whereby parentheses serve to make clear the *scope* of negation signs—and other connectives. Thus, the negation of the conjunction

$$P \wedge Q$$

would be written as follows:

$$\neg(P \wedge Q)$$

If, in contrast, we intend the negation sign to apply exclusively to the first conjunct of a sentence, we need only omit the parentheses

$$\neg P \wedge Q$$

In the case of negation, we employ the following rule: *A negation sign applies only to the expression to its immediate right.* Consider the following sentences:

$$\neg(P \wedge Q) \wedge R$$
$$\neg((P \wedge Q) \wedge R)$$
$$P \wedge (Q \wedge \neg R)$$

In the first sentence, the scope of the negation sign includes the conjunction $(P \wedge Q)$; it stops short of the right conjunct, R. In the second sentence, however, the entire complex expression is negated. In the last sentence, the scope of the negation sign includes only the rightmost conjunct, R.

Exercises 2.01

Provide L_s translations of the English sentences that
follow using the ¬ and ∧ connectives. Let E = Elvis
croons; F = Fenton investigates; G = George flees; H =
Homer flees.

1. George and Homer flee.

2. Homer flees and George flees.

3. Fenton investigates and Homer doesn't flee.

4. It's not the case that both Fenton investigates
    and Homer doesn't flee.

5. Fenton investigates, Elvis croons, and George flees.

6. It's not the case that George and Homer flee.

7. Homer and George flee, and Fenton doesn't
    investigate.

8. Homer and George don't flee.

9. Fenton doesn't investigate, and George and Homer
    don't flee.

10. It's not the case that Homer and George flee, and
    Fenton doesn't investigate.

2.06 DISJUNCTION: ∨

The third truth functional connective to be defined for L_s encompasses
disjunction, symbolized by a wedge, ∨. In English, disjunction is most
familiarly expressed by the phrase "either…or…," as in the sentence

Either it's raining or the sun is shining.

Disjunction in $\mathcal{L}_s$ differs in certain important respects from its English counterpart. The differences become clear once we provide a truth table characterization of the connective:

p	q	$p \vee q$
T	T	T
T	F	T
F	T	T
F	F	F

As the truth table indicates, a disjunction in $\mathcal{L}_s$ is false when, and only when, *both* of its constituent sentences (or *disjuncts*) are false; it is true otherwise.

The second, third, and fourth rows of the truth table coincide nicely with our understanding of disjunction in English. Suppose I proclaim the disjunction above, "Either it's raining or the sun is shining." You would regard my utterance as true if one of the disjuncts is true, if it is raining but the sun isn't shining (the situation depicted in the second row of the truth table) or if the sun is shining and it is not raining (the third row). Similarly, you would take my utterance to be false if it turned out to be false that the sun is shining and false that it is raining (the table's fourth row).

It is harder to square the first row of the truth table characterization with typical English usage. According to the first row, if *both* sides of a disjunction are true, then the disjunction *as a whole* is true. This might seem at odds with English usage. If, for example, you say to me

Iola will arrive Monday or Tuesday

I would expect to greet Iola on Monday *or* on Tuesday, but *not* on *both* days. This may well be the most common way of understanding "either...or..." constructions in English.

Were we to spell out what we have in mind when we use a sentence like that above, we might put it like this:

Iola will arrive on Monday or on Tuesday,
but not on both Monday and Tuesday.

Constructions of this sort express *exclusive disjunction*: "either...or..., *not both*." As the truth table makes plain, a disjunction in $\mathcal{L}_s$ is true if *both* of its disjuncts are true: "either...or..., *maybe both*." While such constructions —*inclusive disjunctions*—occur less frequently in English than exclusive disjunctions, they do occur. Consider the sentence

Employees will be paid time-and-a-half for
working either on weekends or on holidays.

This sentence might appear on a contract you make with your employer. It
means, of course, that you will be paid extra for work done outside normal
working hours: on weekends, holidays—or *both*. You would not look kindly
on an employer who insisted on interpreting the disjunctive clause in an
exclusive way, and refusing, for instance, to pay you time-and-a-half for the
hours you put in last Saturday on the grounds that last Saturday was the
fourth of July, and it is false that time-and-a-half need be paid for work on
days that fall *both* on weekends and on holidays.

We must distinguish, then, *exclusive* disjunction, *either...or, not both*,
from L_s-style *inclusive* disjunction, *either...or, maybe both*. You might think
it odd that we should elect to define disjunction in L_s in this inclusive vein,
but the reason is simple: by using ∨ to represent inclusive disjunction, we
can have our cake and eat it. We can, with a little ingenuity, construct
sentences that express inclusive disjunctions as well as sentences that express
exclusive disjunctions.

Ambiguous Sentences

Some of the sentences in the previous exercises are
ambiguous, they have more than one meaning, hence
more than one translation. Try to identify the sentences
that are ambiguous. Notice whether distinct
translations into L_s are required depending on which
meaning is selected.

Suppose we interpret the L_s sentence, W, as "Socrates is wise," H as
"Socrates is happy," and B as "Socrates is bored." Given these
interpretations, together with our understanding of the truth functional
connectives ¬, ∧, and ∨, we are in a position to produce limitless complex L_s
sentences. Consider the following together with their English equivalents:

$$(W \wedge B) \vee H$$
Socrates is wise and bored, or he is happy.

$$(\neg W \vee B) \wedge \neg H$$
Socrates isn't wise or he's bored, and he isn't happy.

$$\neg W \lor \neg H$$

Socrates isn't wise or he isn't happy.

Provided we recognize that the third sentence leaves open the possibility that Socrates is *both* unwise and unhappy, these translations are straightforward.

A glance at the sentences above reveals a simple way of representing *exclusive disjunctions in* L_s. Return for moment to the sentence

Iola will arrive Monday or Tuesday.

This sentence, we noted, would most naturally be used to assert that Iola will arrive on Monday or on Tuesday, but not on both Monday and Tuesday. Suppose we break this asserted content down into components. First, the sentence informs us that Iola will arrive on either Monday or Tuesday—and not, say, on Friday. This might be expressed as follows:

$$M \lor T$$

The sentence *also* informs us that Iola will *not* arrive on *both* Monday *and* Tuesday. (At least it so informs us, *given* background information—for instance, that an *arrival* occurs at the onset of a visit, and that Iola's visit can begin on Monday only if it does not begin on Tuesday, and vice versa.) This aspect of the sentence's meaning can be captured in L_s by using a negated conjunction

$$\neg(M \land T)$$

This negated conjunction asserts that it is not the case that Iola will arrive both on Monday *and* on Tuesday. The negation sign includes within its scope the *entire* conjunction.

A negated conjunction differs crucially from a conjunction of negations. The sentence above means something quite different from

$$\neg M \land \neg T$$

This sentence means that Iola will not arrive on Monday *and* not arrive on Tuesday, that she will arrive on *neither* day. That is patently *inconsistent* both with the disjunctive sentence with which we began, "Iola will arrive either Monday or Tuesday," *and* with the L_s sentence

$$M \lor T$$

that we have already agreed captures at least a part of the meaning of the English original. *No* disjunction is true if both of its disjuncts are false.

Ambiguity in $\mathcal{L}_s$

$\mathcal{L}_s$ makes use of the familiar mathematical convention of using parentheses to *disambiguate* sentences. (An ambiguous sentence is disambiguated when its intended meaning is made clear.) The expression

$$A \supset B \wedge C$$

is ambiguous as between

$$(A \supset B) \wedge C$$

and

$$A \supset (B \wedge C)$$

$\mathcal{L}_s$ can contain no ambiguous sentences, so the first expression is not a sentence of $\mathcal{L}_s$. It can be turned into a sentence by the addition of parentheses in either of the two ways set out above. Do these two sentences have the same meaning? For practice construct a truth table for each and compare their truth conditions.

By now it should be clear that the exclusive disjunctive sense of our original English sentence can be represented in $\mathcal{L}_s$ by means of a conjunction of $\mathcal{L}_s$ sentences

$$M \vee T$$

which introduces the "either...or..." component of the English original, and

$$\neg(M \wedge T)$$

which introduces the "...not both" component. When conjoined, these two segments become

$$(M \vee T) \wedge \neg(M \wedge T)$$

The first conjunct of this complex expression captures the *disjunctive* aspect of the English sentence, while its second conjunct captures its *exclusive* aspect.

In practice we can adopt the following principle in translating disjunctions from English into $\mathcal{L}_s$: translate disjunctions as *inclusive* disjunctions (that is, just using the $\vee$) unless they are only interpretable as exclusive disjunctions. If a sentence *could* be read as expressing an inclusive disjunction, even if it *might* be read as expressing an exclusive disjunction as well, it should be translated by means of the $\vee$ alone.

Exercises 2.02

Provide $\mathcal{L}_s$ translations of the English sentences that follow using the $\neg$, $\wedge$, and $\vee$ connectives. Let E = Elvis croons; F = Fenton investigates; G = George flees; H = Homer flees.

1. Elvis croons and Homer flees.

2. Homer flees or George doesn't.

3. Fenton investigates and either Homer doesn't flee or George flees.

4. It's not the case that either Fenton investigates or Homer doesn't flee.

5. Either it's not the case that Fenton investigates or Homer doesn't flee.

6. Either Homer or George flees, but not both.

7. Elvis croons and Homer and George flee, or Fenton doesn't investigate.

8. Homer or George doesn't flee.

9. Fenton doesn't investigate or Elvis doesn't croon, but not both.

10. Either Elvis croons or George or Homer flees.

2.07 THE CONDITIONAL: ⊃

The fourth truth functional $\mathcal{L}_s$ connective is the *conditional.* The horseshoe symbol, ⊃, as characterized in the truth table below, expresses in $\mathcal{L}_s$ roughly what is expressed in English by the phrases "…only if…" and "if…then…." Conditional sentences have a central role in $\mathcal{L}_s$, one best appreciated if we compare them to "if…then…" sentences in English. Consider, first, the truth table characterization of the ⊃.

p	q	$p \supset q$
T	T	T
T	F	F
F	T	T
F	F	T

Reading $p \supset q$ as "p is true *only if* q is true" or "*if* p is true, *then* q is true," let us see how close conditionals in $\mathcal{L}_s$ come to those in English. The first two rows of the truth table fit nicely with our pre-$\mathcal{L}_s$ conception of the conditionality. A conditional assertion is true if both its *antecedent* (the sentence at the left) and its *consequent* (the sentence at the right) are true. Similarly, a conditional sentence with a true antecedent and a false consequent is clearly false. Consider, for instance, the English sentences

It's raining only if the street is wet

and

If it's raining then the street is wet.

Both of these sentences can be translated into $\mathcal{L}_s$ as

$$R \supset W$$

If I assert such a conditional, you are unlikely to object if you notice both that it is raining and that the street is wet, that is, if you notice that both its antecedent and its consequent are true. This corresponds to the first row of the truth table characterization. If you notice that it is raining (that the antecedent of the conditional is true) and that the street is not wet (its consequent is false), you would declare my conditional false, and your so doing would correspond to the truth table's second row.

Now imagine a situation in which your observations correspond to the third and fourth rows of the truth table. Suppose, for instance, you observe that it is not raining but that the street is nevertheless wet (the truth table's third row), or that it is not raining and the street is dry (the last row). Is it obvious that in those cases you ought to regard what I have said as *true*?

A short, but unsatisfying answer to this question is that since these observations do not show my conditional sentence to be *false*, they show it to be true. Sentences in L_s must be either true or false, so an L_s sentence that is not false is thereby true. Perhaps this is not so for English sentences. Your observing a cloudless sky and a dry street does not show my sentence to be false, but your observation seems not to show that it is *true* either. Perhaps English conditionals are *neither* true nor false when their antecedents are false, perhaps under those circumstances they lack a truth value. Although this is a possibility we cannot ignore, we should first determine whether there might be simpler explanations for the apparent lack of fit between conditionals in English and those in L_s.

Truth and Truth Conditions

Every sentence of L_s has a set of *truth conditions*: those circumstances under which it is true, and those circumstances under which it is false.

Every sentence of L_s has, as well, a *truth value*: it is either *true* or *false*.

The truth value of a sentence is fixed by:

1. the sentence's truth conditions;

2. the state of the world.

You can know the truth conditions of a sentence without knowing its truth value. You know the truth conditions of the sentence "There is a pound of gold within one mile of the north pole of Venus," even though you *do not know* its truth value, whether it is true or false.

Let us back up and think more carefully about the logic of ordinary English conditionals. For the time being, we can stick with our simple example

If it's raining then the street is wet

which is, I have suggested, equivalent to "It's raining only if the street is wet" and to the L_s sentence

$$R \supset W$$

Consider one obvious feature of this conditional: in uttering it, I need not be interested in whether it is, at the time, raining, or whether the street happens to be wet. My utterance is not intended to communicate the current state of the world. My aim, rather, is to assert that *it is false that it is both raining and the street is not wet.* This is explicit in the second row of the truth table characterization of the $\supset$.

Suppose we put this gloss on my sentence into L_s

$$\neg(R \wedge \neg W)$$

and suppose we agree that this conjunction reasonably captures what I had in mind in asserting the original conditional sentence. Where does this leave us with respect to our original characterization of conditionals in L_s? We seem to have shown that the English sentence

It's raining only if the street is wet

means the same as the sentence

It's not the case that it is raining and the street isn't wet.

But what does it mean to say that two sentences *mean the same?*

Consider what you know when you know what sentences like those above mean. Whatever else you may know, you know the conditions under which the sentence in question is true or false, you know the sentence's *truth conditions.* This does not mean that you know *whether* the sentence is true or false. You know the truth conditions of the sentence "At this moment, the number of pigeons on the Capital dome is even," although you have no idea whether the sentence is true or false. Let us say, then, that the meaning of a sentence is connected in some important way with its truth conditions. At any rate, two sentences with the same meaning might be thought to have the same truth conditions. In translating from one language to another, for instance, we seek sentences that share truth conditions.

In working out the truth conditions for ordinary English sentences, we are obliged to fall back on our tacit, intuitive knowledge of the language. In

L_s matters are different. L_s is a truth functional language. We possess precise characterizations of its truth functional connectives. As a result, we can easily specify the truth conditions of any L_s sentence. To do so, we construct a truth table for the sentence in question. Thus far we have used truth tables only to provide formal characterizations of connectives. Given these characterizations, however, we can devise truth table *analyses* of particular L_s sentences, analyses that provide a clear specification of the truth conditions of any sentence expressible in L_s. Consider the L_s sentence

$$R \supset W$$

It is easy to construct for this sentence a truth table that makes its truth conditions explicit. To do so, we need only call to mind our earlier truth table characterization of the $\supset$ connective:

p	q	$p \supset q$
T	T	T
T	F	F
F	T	T
F	F	T

This truth table tells us, in essence, that a conditional sentence in L_s is true except when its antecedent (that is, its left-hand constituent) is true and its consequent (what is to the right) is false, the situation realized in the second row of the truth table above.

We can apply this information to our L_s conditional so as to yield the following truth table:

R	W	$R \supset W$
T	T	T
T	F	F
F	T	T
F	F	T

We know that the sentences R and W can each be true or false and that, in consequence, there are four possible truth value combinations of these sentences to consider. The truth table characterization of the $\supset$ informs us that an L_s conditional is false when its antecedent is true and its consequent is false; it is true otherwise. Putting all this together, we arrive at the truth table above.

We can think of this truth table as explicitly setting out the truth conditions of our original sentence. You could be said to *know* its truth conditions so long as you can reconstruct its truth table. We have seen that knowing the truth conditions of a given sentence is not the same as knowing its truth *value*. Whether a sentence is true depends on which row of the truth table is *realized*, that is, how things stand in the world.

Possible Worlds

The notion of possible worlds originated with the philosopher G. W. Leibniz (1646–1716). Leibniz noted that the actual world might have been different in countless ways. Each of these ways it might have been is a possible world.

Leibniz famously argued we can account for the existence of the actual world only by supposing that the actual world is the *best possible* world—prompting a cynical response from Voltaire (1694–1778): "If this is the best of all possible worlds, I should hate to see the others!"

Do possible worlds (other than the actual world) exist? Some philosophers have thought that they *do*, on the grounds that *there are*, objectively, ways the actual world might have been. Others regard talk about possible worlds as a convenient fiction.

We can think of each row of a truth table as describing a set of *possible worlds*—where a "possible world" is simply *a way the world might be.* Knowing the truth conditions for a given sentence is a matter of knowing its truth value in *all possible worlds*, knowing what its truth value *would* be were the world a particular way. That might seem a daunting task, but the appearance is misleading. There is an infinitude of possible worlds, but only four possible combinations of truth values for the sentences R and W. What the truth table tells us is that in any world—that is, under any circumstances—in which R is true and W is false, the conditional $R \supset W$ is false. In every other world, it is true. If you know the truth value of a given sentence in addition to its truth conditions, then you know which of these sets of possible worlds includes the actual world.

Let us return now to the L_s sentence we agreed came close to capturing what we have in mind in asserting an ordinary English conditional. We began with the sentence

> If it's raining then the street is wet

and decided that this sentence had the same truth conditions as the sentence

> It's not the case that it is raining and the street isn't wet

a sentence we translated into L_s as

$$\neg(R \wedge \neg W)$$

Let us work up a truth table for this sentence, inspect its truth conditions, and compare these with the truth conditions of the conditional sentence. To accomplish this, we need first to break down the sentence into its constituent sentences. That is, before we determine the truth conditions for the sentence as a whole, we must determine the truth conditions for

$$R \wedge \neg W$$

the sentence contained inside the parentheses. Before working out the truth conditions for *this* sentence, we must calculate the truth conditions for its right-hand conjunct

$$\neg W$$

Armed with that information, we can construct a truth table for the entire sentence.

This step-by-step procedure is set out in the truth table below:

R W	$\neg W$	$R \wedge \neg W$	$\neg(R \wedge \neg W)$
T T	F	F	T
T F	T	T	F
F T	F	F	T
F F	T	F	T

The two left-most columns provide an inventory of the possible combinations of the truth values of R and W. The next column sets out the truth conditions for $\neg W$, that is, the truth value of $\neg W$ given particular truth values for R and W. From the truth table definition of $\neg$, we know that $\neg W$ is true when W is false, and false when W is true. Now, relying on our truth table characterization of $\wedge$, we can specify truth conditions for the conjunction $R \wedge \neg W$. A conjunction is true only when both of its conjuncts are true, and false otherwise. So the conjunction $R \wedge \neg W$ is true, only in those rows of the truth table (only in those possible worlds) in which both R and $\neg W$ are true. The relevant columns are the R column and the $\neg W$ column.

Having established the truth conditions for the conjunction $R \wedge \neg W$, we need only determine the truth conditions for the negation of this conjunction. Recalling our characterization of negation in $\mathcal{L}_s$, we know that negating a sentence reverses its truth value: a true sentence, negated, is false, and the negation of a false sentence is true. Thus whenever the sentence $R \wedge \neg W$ is true, its negation, $\neg(R \wedge \neg W)$ is false, and whenever $R \wedge \neg W$ is false, its negation is true. This is reflected in the right-most column of the truth table.

The truth table we have just constructed provides an explicit representation of the truth conditions of the $\mathcal{L}_s$ sentence we agreed came close to capturing the sense of our original English conditional. If we now compare that truth table with the truth table we constructed for the corresponding $\mathcal{L}_s$ conditional, we can see there is a perfect match:

R W	$\neg(R \wedge \neg W)$	$R \supset W$
T T	T	T
T F	F	F
F T	T	T
F F	T	T

The truth table shows clearly that the truth conditions for the two $\mathcal{L}_s$ sentences are identical so that, for our purposes, they mean the same. Given that the one approximates our English original, the other must as well.

All this suggests that $\mathcal{L}_s$ conditionals are closer in meaning to English "if...then..." sentences than is usually supposed. It does not follow that conditional constructions in English invariably express conditionals in this sense. English sentences containing "if...then..." clauses may be used to express other relations. At first this may lead to intermittent fits of anxiety when we set out to translate English sentences into $\mathcal{L}_s$. Occasionally, English sentences that *look* like $\mathcal{L}_s$-style conditionals will turn out not to be conditionals at all. Eventually, if we persist, these difficulties will sort themselves out and translation will seem almost natural.

2.08 CONDITIONALS, DEPENDENCE, AND SENTENTIAL PUNCTUATION

One further feature of our treatment of conditional sentences bears mention. In describing a sentence of the form "If it's raining, then the street is wet" as a conditional sentence, it is natural to regard the consequent, "The street is wet," as *conditional* or *dependent* on the antecedent, "It's raining." This is unobjectionable so long as the dependence in question is recognized to be *logical* rather than *causal*. The distinction is easy to miss. Logical and causal dependencies often go hand in hand in English discourse. Consider the sentence

If you drink epoxy, then you'll become ill.

When we hear this sentence, we immediately envisage a causal connection between your drinking epoxy and your consequent illness. Causes precede effects, so we may understand the sentence to describe a sequence in which drinking epoxy is followed by illness.

We have seen that conditional "if...then..." sentences are equivalent to "...only if..." sentences. The sentence above might be paraphrased as

You drink epoxy only if you become ill.

This paraphrase *looks* wrong. It appears to reverse the order of dependence: your becoming ill now seems to lead to your drinking epoxy!

The appearance is misleading. It stems from a tendency to run together logical and causal relations. The original sentence expresses, among other things, a particular logical relation between the sentences "You drink epoxy" and "You become ill": if the first is true, then the second is true as well. Alternatively, it is not the case that the first is true and the second false. It might be that what makes this sentence true is the holding of a causal relation between your drinking epoxy and your becoming ill. But we can consider logical relations sentences express independently of whatever it is about the world that makes those sentences true or false. Conditional constructions compel us to do precisely that.

In practice, we can accustom ourselves to distinguishing logical from causal relations by focusing explicitly on the truth conditions of sentences. Faced with the sentence, "If you drink epoxy, then you'll become ill," we notice first that the sentence comprises two component sentences, "You drink epoxy" and "You become ill." Second, we puzzle out the logical relations holding between these two sentences, that is, we endeavor to tease out the relation between the truth conditions of the original conditional sentence and those of the sentences that make up that sentence's antecedent and consequent. We recognize that the conditional sentence is false when (1) the sentence "You drink epoxy" is true *and* (2) the sentence "You become ill" is false: the first sentence is true *only if* the second is true. This is just what is captured in $\mathcal{L}_s$ by the $\supset$ connective. The English sentence doubtless conveys *more* than this logical relation, but it expresses at least this much.

In translating English "if...then..." conditionals into $\mathcal{L}_s$, you should begin by first locating the if-clause. The sentence associated with the if-clause goes to the left of the conditional sign when the sentence is translated into $\mathcal{L}_s$. Although this might seem obvious, it is not so obvious when a sentence's if-clause is buried in the middle of the sentence, or when it follows a consequent then-clause. We can say

> If it's raining then the street is wet

or

> The street is wet if it's raining.

Both sentences are translated into $\mathcal{L}_s$ as

$$R \supset W$$

In translating the second sentence, we locate the if-clause, "It's raining," and identify this as the antecedent of the conditional of which it is an element.

If-clauses are one thing, *only if*-clauses are another. A sentence of the form

$$p, \text{ if } q$$

is translated into $\mathcal{L}_s$ as

$$q \supset p$$

In contrast, a sentence of the form

$$p \text{ only if } q$$

goes into $\mathcal{L}_s$ as

$$p \supset q$$

Consider the English sentence

> Gertrude will leave *if* Fenton investigates.

Assuming that G = "Gertrude will leave," and F = "Fenton investigates," this sentence can be translated into $\mathcal{L}_s$ as

$$F \supset G$$

This sentence is very different from the superficially similar sentence

> Gertrude will leave *only if* Fenton investigates

as its $\mathcal{L}_s$ translation reveals

$$G \supset F$$

In confronting an English conditional, then first look for an if-clause. This will be the antecedent of the conditional—*unless* the clause is an only-if-clause. In that case, the clause embodies the consequent of a conditional.

Exercises 2.03

Translate the following sentences into L_s using the $\neg$, $\wedge$, $\vee$, and $\supset$ connectives. If the English sentence is ambiguous, provide distinct L_s versions of it. But remember: no L_s sentence can be ambiguous. Let E = Elvis croons; F = Fenton investigates; G = George flees; H = Homer flees.

1. If Elvis croons, Homer flees.

2. Homer flees if Elvis croons.

3. Homer flees only if Elvis croons.

4. If George or Homer flees, Fenton investigates.

5. Fenton investigates if George or Homer flees.

6. If Elvis croons, then, if George flees, Fenton investigates.

7. If George and Homer flee, Elvis doesn't croon.

8. Homer or George flees if Elvis croons.

9. Fenton doesn't investigate if Elvis doesn't croon.

10. Fenton investigates only if either Elvis croons or George or Homer flees.

2.09 THE BICONDITIONAL: $\equiv$

Once we grasp the logic of conditional sentences, biconditionals are easy. Biconditional sentences resemble two-way conditionals. English biconditionals are associated with the phrases "...*if and only if*..." and "...*just in case*...." Consider the English sentence

> This solution is acid if and only if it turns litmus paper red.

This sentence incorporates the truth conditions not only of the conditional

> If this solution is acid, then it turns litmus paper red

but also of the complementary conditional

> If the solution turns litmus paper red, then it is acid.

Biconditionals in this way express bidirectional, "back-to-back" conditionals in a single sentence. For this reason, biconditionals can be paraphrased by conjoining a pair of complementary conditionals.

Consider another English example

> A number is prime just in case it is divisible only by itself and one.

The sentence can be paraphrased by means of the conjoined pair

> If a number is prime, then it is divisible only by itself and one

and

> If a number is divisible only by itself and one, then it is prime.

Ordinary conditionals are not in this way bidirectional. For instance, the English conditional

> If it's Ferguson's bread, then it's good

does not imply

> If it's good, then it's Ferguson's bread.

The truth of the original conditional sentence is compatible with the falsehood of the sentence above. I might assert with perfect consistency that if it's Ferguson's bread then it's good, while denying that if it's good, then it's Ferguson's bread: plenty of things other than Ferguson's bread are good.

These remarks about English biconditionals apply straightforwardly to biconditionals in L_s. Suppose we construct a truth table for the conjoined conditional pair

$$(A \supset R) \wedge (R \supset A)$$

and compare the truth conditions of this sentence to those for the corresponding biconditional.

A R	$A \supset R$	$R \supset A$	$(A \supset R) \wedge (R \supset A)$
T T	T	T	T
T F	F	T	F
F T	T	F	F
F F	T	T	T

The table below provides a characterization of the biconditional in $\mathcal{L}_s$.

p q	$p \equiv q$
T T	T
T F	F
F T	F
F F	T

Appealing to this characterization, we can set out the truth table for the biconditional corresponding to our original English sentence

A R	$A \equiv R$
T T	T
T F	F
F T	F
F F	T

If we compare the truth conditions of this sentence with those of the conjoined (back-to-back) conditionals above, we can see that they are identical. The sentences, despite their different appearances, have the same truth conditions, hence, for our purposes, they mean the same. We agreed that English biconditionals were nicely captured using conjoined, back-to-back conditionals in $\mathcal{L}_s$. Biconditionals in $\mathcal{L}_s$ have the same truth conditions as back-to-back $\mathcal{L}_s$ conditionals. We are thereby warranted in supposing that the biconditional in $\mathcal{L}_s$ parallels biconditionality in English.

┌─────────────── **Exercises 2.04** ───────────────┐

Translate the following sentences into $\mathcal{L}_s$. If the English
sentence is ambiguous, provide distinct $\mathcal{L}_s$ versions of it.
But remember: no $\mathcal{L}_s$ sentence can be ambiguous. Let E
= Elvis croons; F = Fenton investigates; G = George
flees; H = Homer flees.

1. Homer flees if Elvis croons,

2. Homer flees only if Elvis croons.

3. Homer flees if and only if Elvis croons.

4. Homer flees if and only if either Elvis croons or
   Fenton Investigates.

5. Elvis croons just in case George or Homer flees.

6. Elvis croons just in case George and Homer flee.

7. Elvis croons if and only if George and Homer don't
   flee or Fenton doesn't investigate.

8. Homer or George flees just in case Elvis croons.

9. If Fenton doesn't investigate then Elvis doesn't
   croon if and only if George doesn't flee.

10. If Fenton investigates then Elvis croons just in
    case George doesn't flee.

└──┘

2.10 COMPLEX TRUTH TABLES

We have seen that truth tables can be used both to characterize truth
functional connectives and to assess the truth conditions of $\mathcal{L}_s$ sentences.
These uses are related in an obvious way. The truth conditions of a complex
sentence can be calculated only if we already know the formal
characterizations of the connectives it contains. Let us look briefly at some

general principles useful in the construction of truth tables for complex $\mathcal{L}_s$ sentences.

We have seen that the truth table for a sentence provides an account of that sentence's truth conditions. It does this by exhibiting the truth value of the sentence in all possible worlds: its truth value under relevantly different circumstances. These circumstances or possible worlds are represented by rows on the truth table. Most sentences will be true in many possible worlds and false in many others, so we take each row to represent a *set* of possible worlds, the set, namely, in which the sentences we are charting have a particular assignment of truth values. Given an arbitrary sentence, S, we might wonder *how many* sets of possible worlds we ought to consider, how many rows we must include in S's truth table.

Happily, there is a simple way to calculate the number of rows required. First, we count the number of *distinct atomic sentences* contained in the target sentence. Every $\mathcal{L}_s$ sentence contains at least one atomic constituent, and it can contain many more. The number of distinct atomic sentences is arrived at by counting occurrences of distinct uppercase letters. Thus, the sentence

$$\neg P \supset Q$$

contains *two* distinct uppercase letters, P and Q, and the sentence

$$\neg P \supset (Q \vee \neg R)$$

contains *three*, P, Q, and R. Notice that the sentence

$$\neg P \supset (Q \vee (\neg R \wedge P))$$

contains just *three* distinct uppercase letters. There are four uppercase letters, but only three *distinct* uppercase letters: P, Q, and R.

Given this number, we can calculate how many sets of possible worlds we shall have to consider, how many rows to include in our truth table. We know that each atomic sentence has one of two values: true or false. Our truth table must provide an inventory of *all possible combinations* of these truth values. Each sentence can have one of two values, so there will be 2^n possible combinations of values, where n is the number of distinct atomic sentences. This means that truth tables for sentences containing a single atomic sentence require two rows ($2^1 = 2$), truth tables for sentences containing two distinct atomic constituents require four rows ($2^2 = 4$), those with three distinct constituents require eight rows ($2^3 = 8$), and so on.

In addition to making sure we have the correct number of rows, we must also be certain to include in a truth table every possible *combination* of truth values, every relevant set of possible worlds. A truth table may have enough rows but fail to do this because some rows repeat others. Were that to happen, we would have left out one or more possible truth value combinations, and our truth table would be defective. To insure that every possible combination of truth values occurs, let us adopt the convention of constructing truth tables according to a simple pattern. Consider an $\mathcal{L}_s$ sentence introduced already

$$\neg P \supset (Q \vee (\neg R \wedge P))$$

We have noted that this sentence contains three distinct atomic constituent sentences, so we know that our truth table will require *eight* rows reflecting the eight possible combinations of truth values of P, Q, and R. The convention for assigning truth values to truth table rows is this:

1. Set out at the left of the table the atomic constituents of the sentence for which the truth table is being constructed.

2. Determine the number of rows required in the manner explained above. The truth table will require 2^n rows for n distinct atomic sentences.

3. In the right-most column, in this example, the R-column, *alternate* Ts and Fs, beginning with Ts.

4. In the next column to the left, here the Q-column, again starting with Ts, alternate *pairs* of Ts and Fs.

5. In the next column to the left, here the P-column, alternate *quadruples* of Ts and Fs, beginning with Ts. Continue this procedure until the left-most column is reached, each time doubling the length

of strings of *T*s and *F*s, and alternating these groups beginning with *T*s. In the left-most column, the first $2^n/2$ rows will contain *T*s, and the remaining $2^n/2$ rows will hold *F*s.

P	Q	R	
T	T	T	
T	T	F	
T	F	T	
T	F	F	
F	T	T	
F	T	F	
F	F	T	
F	F	F	

This technique, in addition to assuring that we have an exhaustive listing of truth value combinations, also insures that truth tables exhibit a *standard pattern* that makes our use of them much less tedious than it might be otherwise.

Having established a framework, we can move on to calculate the truth conditions of the target sentence. It is easy, in the case of complex sentences, to make mistakes. Because a mistake at one point is inherited elsewhere, it is wise to proceed cautiously and systematically. Rather than attempting to ascertain the truth conditions for a complex sentence *all at once*, it is safer to work out the truth conditions for its components, moving "inside-out" from smaller to larger units until the level of the whole sentence is reached. In practice, the procedure works as follows:

1. Calculate the truth conditions for every *negated atomic sentence*. The target sentence contains *two* negated atomic sentences, $\neg P$ and $\neg R$.

2. Calculate the truth conditions of sentences within parentheses working from the innermost parentheses out. In our example, this means that we calculate the truth conditions for $(\neg R \land P)$, then those for $Q \lor (\neg R \land P)$. In this way we arrive eventually at the truth conditions for the whole sentence.

This procedure is illustrated in the truth table below. Truth conditions for $\neg P$ and $\neg R$ are set out at the left, and those for subsequent sentences make use of these calculations.

$P\ Q\ R$	$\neg P$	$\neg R$	$\neg R \wedge P$	$Q \vee (\neg R \wedge P)$	$\neg P \supset (Q \vee (\neg R \wedge P))$
T T T	F	F	F	T	T
T T F	F	T	T	T	T
T F T	F	F	F	F	T
T F F	F	T	T	T	T
F T T	T	F	F	T	T
F T F	T	T	F	T	T
F F T	T	F	F	F	F
F F F	T	T	F	F	F

Confronted by a sprawling truth table, you might be tempted to take short-cuts. Short-cuts are not advised. The more complex the sentence, the more ways you can go wrong, and the more caution you must exercise to avoid error.

Once the principles of truth table construction are grasped, it is a simple, though occasionally grueling, matter to construct truth tables for any $\mathcal{L}_s$ sentence and thereby to specify in a rigorous way that sentence's truth conditions. Were you to do this long enough and on enough sentences, you would begin to discover sentence pairs that, while differing syntactically, in their arrangement of elements, nevertheless exhibited *identical truth conditions*. We have noted already that in $\mathcal{L}_s$, when two sentences have the same truth conditions, we are entitled to regard them as having the same meaning. None of this is surprising. We say that Iola procrastinates and, when asked to explain, we say that Iola puts things off. In this case, the English sentences

Iola procrastinates

and

Iola puts things off

are *paraphrases* of one another and have the same truth conditions: they will be true in the same possible worlds and false in the same possible worlds. To test this intuition you might try to envisage a possible world in which the sentences possess *different* truth values.

When we discover paraphrase relations in a natural language like English, we discover, in effect, ways we might *eliminate* elements of the language without affecting our ability to describe our world. The sentences above illustrate the point. Given that those sentences are paraphrases, we might eliminate the word "procrastinate" from English, thereby simplifying the language, without affecting its fact-stating capacity. In eliminating terms, of course, we might affect other aspects of the language, for instance, its ability to express facts elegantly or poetically.

Such considerations need not deter us when we turn to $\mathcal{L}_s$. $\mathcal{L}_s$ is a formal idiom with no poetic pretentions. That being the case, you might wonder *how far* the vocabulary of $\mathcal{L}_s$ could be paired down without affecting its ability to express propositions. Perhaps not at all; perhaps we need every element. We need a stock of atomic sentences, so there would be no point in trying to limit these. But what about our cherished truth functional connectives? Might *these* somehow be reduced in number? Were their numbers diminished, what would be the effects on those of us obliged to put $\mathcal{L}_s$ through its paces?

Recall our discussion (§ 2.9) of biconditionals. We discovered that every biconditional sentence corresponds to a pair of back-to-back conditionals. Thus, the sentence

$$P \equiv Q$$

might be paired with the sentence

$$(P \supset Q) \wedge (Q \supset P)$$

These sentences are *logically equivalent*, that is, they have identical truth conditions—as we could easily prove by constructing a truth table for each sentence. The second sentence provides a paraphrase of the first. This suggests a way of *eliminating* the biconditional connective, $\equiv$, from $\mathcal{L}_s$ without affecting the language's expressive capabilities: whenever a biconditional is called for, we could instead use back-to-back conditionals.

It would seem, then, that we could dispense with at least one element of $\mathcal{L}_s$, the biconditional connective. What of the others? Think back to our earlier discussion (§§ 2.7–2.8) of conditionals in $\mathcal{L}_s$ and in English. There we sought to motivate the usual truth table characterization of $\supset$ by arguing that ordinary English conditionals of the form

$$\text{If } P \text{ then } Q$$

could be regarded as having the same truth conditions as sentences of the form

$$\text{It's not the case that } P \text{ and not-}Q.$$

Representing the latter in $\mathcal{L}_s$ as

$$\neg(P \wedge \neg Q)$$

we proceeded to construct a truth table that exhibited its truth conditions. Once this was done, we discovered that this sentence had the very same truth conditions as its conditional counterpart

$$P \supset Q$$

At the time, this information was used to make a case for a particular characterization of the $\supset$ connective. Now we can see that the logical equivalence we uncovered provides us with a technique for dispensing with the $\supset$. The situation parallels that of the biconditional: we could forego the $\supset$, replacing it with a combination of $\neg$s and $\wedge$s. We have seen that we could replace occurrences of $\equiv$ with back-to-back conditionals using just $\wedge$s and $\supset$s; we might do without *both* the $\supset$ and the $\equiv$. We replace $\equiv$s with combinations of $\supset$s and $\wedge$s, and then, in the resulting sentences, we replace occurrences of $\supset$s with $\neg$s and $\wedge$s.

Might we go further? Might we eliminate the $\wedge$, the $\neg$, or the $\vee$? If so, *how far* might we go in streamlining $\mathcal{L}_s$? As it happens, we could get by in $\mathcal{L}_s$ with only negation, $\neg$, and *one* of the connectives $\supset$, $\wedge$, $\vee$. For any sentence we might express using these (or of course $\equiv$), we could substitute a *logically equivalent* sentence, a sentence with identical truth conditions, that did not use that connective. We could, in a fit of logical zeal, scrap all of our connectives save the $\neg$ and, say, the $\vee$. The result would be a no-frills language of austere beauty—or, at any rate, an austere language.

Austere beauty has its price. Were we to undertake to purge $\mathcal{L}_s$ of "superfluous" truth functional connectives, we would create a language much less easy to use. Sentences that we find easy to write in $\mathcal{L}_s$ as it is now constituted would become tedious strings of $\neg$s and $\vee$s. To illustrate, consider the sentence

$$P \equiv (Q \wedge R).$$

Compare the following paraphrase of this sentence using only $\neg$s and $\vee$s:

$$\neg(\neg(\neg P \vee \neg(\neg Q \vee \neg R)) \vee \neg((\neg Q \vee \neg R) \vee \neg P))$$

More complex sentences would require correspondingly more complex paraphrases. I leave it to the reader to construct a truth table proof that the sentences above do indeed have the same truth conditions.

Exercises 2.05

Construct a truth table for each of the L_s sentences below.

1. $\neg P \vee Q$ 2. $\neg P \wedge \neg Q$

3. $\neg P \vee \neg Q$ 4. $\neg(\neg P \vee \neg Q)$

5. $\neg(\neg P \wedge \neg Q)$ 6. $P \supset \neg Q$

7. $P \supset (Q \wedge R)$ 8. $P \supset \neg(Q \wedge \neg R)$

9. $P \wedge (Q \vee R)$ 10. $\neg(\neg P \vee (\neg Q \wedge \neg R))$

2.11 THE SHEFFER STROKE: |

Someone might take all this as a challenge. If we could manage with just two truth functional connectives, might we get by with *even fewer*? Might we somehow preserve the expressive power of L_s wielding only a *single* connective? The answer is no and yes. No, we could not get by with just one of our current stock of L_s connectives. We need the negation connective to make paraphrases work, but negation alone would not allow us to write sentences containing more than a single atomic constituent. We might, however, invent a *new sign*, one that *amalgamated* negation and some other logical operation. The result would be a language as expressive as L_s, sublime after a fashion, but unrelentingly dreary to use.

All this was proved years ago by H. M. Sheffer. Sheffer defined a stroke-function (now called the Sheffer-stroke), |, the truth table characterization of which is as follows:

p	q	$p \mid q$
T	T	F
T	F	F
F	T	F
F	F	T

Here $p \mid q$ might be read as "neither p nor q." Surprisingly, it can be shown that all of the truth functional connectives used in $\mathcal{L}_s$ can be replaced by just this one. Imagine a Sheffer-stroke equivalent of the placid $\mathcal{L}_s$ sentence

$$A \wedge B$$

What would it look like? Consider, first, that a simple negation, $\neg p$, expressed via Sheffer-stroke notation, comes out as

$$p \mid p$$

This can be shown by means of a truth table

p	q	$p \mid q$	$p \mid p$
T	T	F	F
T	F	F	F
F	T	F	T
F	F	T	T

Consider now the expression

$$(A \mid A) \mid (B \mid B)$$

and its truth table

A	B	$A \mid A$	$B \mid B$	$(A \mid A) \mid (B \mid B)$
T	T	F	F	T
T	F	F	T	F
F	T	T	F	F
F	F	T	T	F

The truth conditions for this sentence are no different from those for our original conjunction; the sentences are logically equivalent.

These truth tables show that we can capture negation and conjunction by means of Sheffer-stroke notation. We knew already that every $\mathcal{L}_s$ sentence is logically equivalent to a sentence containing only $\neg$s and $\wedge$s. It must be possible, therefore, to construct a Sheffer-stroke equivalent of *any* $\mathcal{L}_s$ sentence. We might, in this way, pair down the logical vocabulary of $\mathcal{L}_s$, although in so doing we should lose simplicity of expression. Short $\mathcal{L}_s$

sentences would become more ungainly and less attractive. The Sheffer-stroke equivalent of

$$A$$

for instance, would be

$$(A \mid A) \mid (A \mid A).$$

For most of us, this is nothing if not confusing.

Alarmed about the prospect of confusing strings of symbols, we might consider *expanding* the set of truth functional connectives—adding, for instance, a connective signifying exclusive disjunction ("either p or q, and not both p and q"). Were we to do so, we would shorten particular L_s sentences, just as we could shorten English sentences by adding terms to the language that replace whole phrases. But adding connectives, like adding terms, has a price. An additional element—together with its effects on the truth conditions of sentences in which it figured—would need to be remembered. On the whole, it is not worth the bother.

Such issues concern, not the expressive power of the language, but the ease with which we use it, its perceived simplicity. L_s has been constructed so as to strike a reasonable balance between notational simplicity and ease of use. It reflects logical notions that in natural languages have evolved distinctive modes of expression. We might, in English, get by without explicit conditional constructions, for instance, just as we can get by with them in L_s: by replacing conditionals with combinations of other elements. But English, like other natural languages, has evolved as it has because it fits us. Similarly, formal languages like L_s have evolved as they have because they mirror logically central features of natural languages.

Exercises 2.06

For each of the $\mathcal{L}_s$ sentences below, devise a logically equivalent sentence, a sentence with the same truth conditions, but one that uses only the truth functional connectives indicated. For each sentence construct a truth table that proves the equivalence.

1. $P \land Q$ $\{\neg, \lor\}$

2. $P \supset Q$ $\{\neg, \land\}$

3. $P \lor Q$ $\{\neg, \supset\}$

4. $P \lor Q$ $\{\,|\,\}$

5. $P \,|\, Q$ $\{\neg, \lor\}$

6. $P \lor Q$ $\{\neg, \land\}$

7. $P \land Q$ $\{\neg, \supset\}$

8. $P \supset Q$ $\{\neg, \lor\}$

9. $P \supset Q$ $\{\,|\,\}$

10. $P \,|\, Q$ $\{\neg, \supset\}$

2.12 TRANSLATING FROM ENGLISH INTO $\mathcal{L}_s$

Translating an English sentence into $\mathcal{L}_s$ is no different in principle from translating that sentence into another natural language. We know in advance that some of the sentence's features may prove untranslatable. Our aim is to find a sentence in the target language with truth conditions matching those of the original sentence. In the case of natural languages, we may wish to preserve, as well, poetic qualities, connotations, and "feel." Owing to cultural differences, these often fail to survive translation or they survive only in some transmuted form. There is, however, no reason to think that we should ever encounter a sentence whose truth conditions could not be matched by *some* sentence or set of sentences in the translator's language.

When it comes to translations into $\mathcal{L}_s$, there is no question of preserving anything other than truth conditions. $\mathcal{L}_s$, although charming in many respects, is stylistically indifferent. More seriously, $\mathcal{L}_s$ lacks logical devices common to all natural languages. English sentences may lack a satisfying $\mathcal{L}_s$ analogue. That will be so when no sentence in $\mathcal{L}_s$ has quite the same truth conditions as the English original. In such cases we are obliged to lower our sights and locate an $\mathcal{L}_s$ sentence whose truth conditions approximate those of the target sentence.

Translatability

Translation is an art, not a science. This is due, in part, to our preference for translations that preserve the flavor of the original. Suppose we focus exclusively on truth conditions. Is there any reason to think that some languages contain sentences whose truth conditions could not be matched by some sentence or set of sentences in our language—English, say, or French, or Urdu?

No one doubts that particular *words* in a language may lack exact counterparts in other languages. Presumably, such words could be paraphrased.

Donald Davidson has argued that languages, whatever their origins, *must* be intertranslatable [see his "The Very Idea of a Conceptual Scheme," in *Inquiries into Truth and Interpretation*, (Oxford: Clarendon Press, 1984)]. Davidson asks: what evidence could we have that an utterance or inscription was a sentence if it could not be translated into a sentence of our language?

The limitations of L_s should not be overstated. L_s meshes nicely with an indispensable segment of English. Difficulties we encounter in constructing translations stem less from the character of L_s than from a tendency to focus on English sentences in a way that obscures their truth conditions. Not that we lack an appreciation of those truth conditions; on the contrary, we all grasp them instantly and unself-consciously. The trick is to turn our implicit knowledge into explicit knowledge. Most of us know how to ride a bicycle and tie a bow. How many of us could make this knowledge explicit? Imagine, for instance, trying to explain to someone how to ride a bicycle or tie a bow over the telephone. You learn your native language by *speaking* it, just as you learn to ride a bicycle or tie a bow by *doing* it—badly, at first, then, with practice, better. Such learning is rarely a matter of our taking in explicit principles or, if it is, only after those principles become second nature do we regard them as having taken hold. In learning to ski, you learn that when you want to turn, you must put your weight on the downhill ski. You master the technique, however, only when you come to do it unthinkingly. In learning a second language, we often learn explicit rules and principles. Only when these become implicit, only when we can "think in the language," can we be said to have learned it.

Speaking or understanding a language, like riding a bicycle, tying a bow, or skiing, is a *skill*. So is translation from English into $\mathcal{L}_s$. The more you do, the easier it is. At first the process seems unnatural, but, with persistence, it can become second nature. At the outset, we must sensitize ourselves to logical features of English sentences that have significant truth conditional roles. These features are familiar to all of us already—otherwise we would not understand one another. But it sometimes requires a struggle to see them for what they are. Most of the rest of this chapter will be devoted to hints and suggestions for bringing the truth conditions of English sentences to the surface. Once we can do that, translation into $\mathcal{L}_s$ is a breeze.

Notice, first, that English word order, although important, is not an infallible guide to the logical character of English sentences. Consider the sentences

> Spot bit Iola.
> Iola bit Spot.

These sentences contain the same words, but in different arrangements. Word order is significant in such cases: it enables us to distinguish the actor and the recipient of the action, or, in classical terminology, the *agent* and *patient*. Do agents inevitably precede patients in sentences of this sort? No. Cases in which the order is reversed are common. Consider, for instance

> Iola was bitten by Spot.

Such examples bring to mind a useful distinction between the *surface structure* of a sentence and its *deep structure*. The sentences

> Spot bit Iola.
> Iola was bitten by Spot.

evidently have the same truth conditions despite differing in word order—and in other ways as well. We might say of such sentences that they are "in one sense" the same and "in another sense" different. Their differences are obvious and outward, detectable even by someone unfamiliar with English. Their sameness is visible only to someone who knows enough English to recognize that the sentences have identical truth conditions. Such sentences differ in their surface structure, but not in their deep structure.

Deep and Surface Structure

The distinction between deep and surface structure is due to the linguist, Noam Chomsky [see, e.g., his *Syntactic Structures*, (The Hague: Mouton, 1957)].

Chomsky argued that in learning a language, we learn the grammar of that language, a set of rules for generating sentences. Some rules generate deep structures; others generate surface structures from these deep structures.

Surface structures are associated with sentences we speak and write. Deep structures are associated with sentences' truth conditions.

The rules that generate surface structures from deep structures include options. Depending on which options are taken, a single deep structure can yield multiple surface structures. When this happens, the resulting sentences will have the same truth conditions. Ambiguous sentences occur when distinct deep structures yield identical surface structures.

Mastering a language involves acquiring the knack of recovering deep forms from surface structures. Our skill at doing this is reflected in the ease with which we recognize the sentences above to have the same meaning. It is reflected, as well, in our recognition that, as in the examples below, a single sentence—a single surface structure—can be associated with distinct truth conditions—distinct deep structures

> Gertrude was frightened by the new methods.
> The lamb was too hot to eat.
> They are flying planes.

Each of these sentences is *ambiguous*, each could be used on different occasions to express different meanings.

Although a grasp of the truth conditions of a sentence requires a grasp of its deep structure, the notions are not equivalent. Deep structure includes abstract elements that combine to determine, *among other things*, a sentence's truth conditions.

The distinction between deep and surface structure is useful to bear in mind as we go about the construction of translations into $\mathcal{L}_s$. Since our aim is to devise $\mathcal{L}_s$ sentences that have the same truth conditions as particular English sentences, and since a sentence's truth conditions are determined by its deep structure, we must beware of being distracted by superficial characteristics of the sentences we translate. The point will be amply and repeatedly illustrated as we consider familiar English constructions.

2.13 CONJUNCTION

The identification of conjunctions in English, and their translation into $\mathcal{L}_s$, is not likely to raise special problems. In the sentences below, for instance, conjunction is easy to spot:

Fenton investigates and Iola procrastinates.
Fenton doesn't investigate and Iola doesn't procrastinate.

Their translation into $\mathcal{L}_s$ is gratifyingly straightforward

$$F \wedge I$$
$$\neg F \wedge \neg I$$

Conjunctions are not invariably signaled in English by the appearance of "and." Each of the sentences below expresses a conjunction. So far as $\mathcal{L}_s$ is concerned each possesses the same truth conditions

Fenton investigates *and* Iola procrastinates.
Fenton investigates *but* Iola procrastinates.
Fenton investigates, *however* Iola procrastinates.
Fenton investigates *yet* Iola procrastinates.
Fenton investigates *although* Iola procrastinates.
Although Fenton investigates, Iola procrastinates.

Each of these sentences goes over into $\mathcal{L}_s$ as

$$F \wedge I$$

The key in translating such sentences is to recognize that despite superficial differences, each expresses a conjunction: each is true just in case each of its conjuncts—"Fenton investigates" and "Iola procrastinates"—is true.

Exercises 2.07

Translate the following sentences into $\mathcal{L}_s$. Let C = Callie escapes; F = Fenton investigates; G = Gertrude investigates; J = Joe is kidnapped.

1. Joe is kidnapped, however Callie escapes,

2. Although Joe is kidnapped, Callie escapes.

3. Fenton investigates but Gertrude doesn't.

4. Fenton and Gertrude don't both investigate.

5. Both Fenton and Gertrude investigate if Joe is kidnapped.

6. Fenton investigates if and only if Joe is kidnapped, but Gertrude doesn't investigate.

7. If Joe is kidnapped and if Callie escapes, then Gertrude investigates.

8. Gertrude investigates if Joe is kidnapped or Callie doesn't escape.

9. Either Fenton investigates and Callie escapes or Joe is kidnapped and Gertrude investigates.

10. Although Gertrude investigates, Fenton doesn't, and Callie escapes.

2.14 DISJUNCTION

Disjunction in $\mathcal{L}_s$ is no more confusing than conjunction. We have already distinguished *exclusive* and *inclusive* disjunctive constructions (see § 2.6). Once this distinction is mastered, the remaining aspects of disjunction are within grasp. In English, disjunction is typically, though not invariably, signalled by an occurrence of "either…or…." Thus, the sentence

Either Fenton investigates or Iola procrastinates

expresses a disjunction, and is translated into $\mathcal{L}_s$ as

$$F \vee I$$

Negated disjunctions are only slightly more complex. Consider the sentence

Neither Fenton nor Iola investigates.

The sentence is a good one on which to reflect in the manner suggested earlier. Suppose we set out to make its truth conditions explicit. In so doing, we can see that the sentence will be true when Fenton does not investigate and Iola does not investigate, and that it is false otherwise—false if Fenton investigates, or Iola investigates, or if they both do. This suggests the following $\mathcal{L}_s$ translation:

$$\neg F \wedge \neg I$$

This translation represents the English original, not as a *dis*junction, but as a *con*junction. If our task is to find an $\mathcal{L}_s$ sentence with the right truth conditions, however, and if this sentence possesses the right truth conditions, then it is a perfectly satisfactory translation.

As it happens, we can also translate the sentence in a way that preserves the disjunctive character of its surface structure. We might, that is, think of the sentence as saying something like

It's not the case that either Fenton or Iola will investigate.

which is translated into $\mathcal{L}_s$ as

$$\neg(F \vee I)$$

If we are right about all this, then both of these $\mathcal{L}_s$ translations should have the same truth conditions. We can check by constructing a truth table:

F I	$\neg F$	$\neg I$	$\neg F \wedge \neg I$	$F \vee I$	$\neg (F \vee I)$
T T	F	F	F	T	F
T F	F	T	F	T	F
F T	T	F	F	T	F
F F	T	T	T	F	T

The truth conditions for the two sentences match—as revealed by the fifth and seventh columns—so they have the same truth conditions. We have uncovered two logically equivalent ways to translate "neither...nor..." sentences.

It might be noted, in passing, that the two $\mathcal{L}_s$ sentences

$$\neg(F \vee I)$$
$$\neg F \vee \neg I$$

are *not* equivalent in meaning—as the truth table below plainly shows:

F I	$\neg F$	$\neg I$	$\neg F \vee \neg I$	$F \vee I$	$\neg (F \vee I)$
T T	F	F	F	T	F
T F	F	T	T	T	F
F T	T	F	T	T	F
F F	T	T	T	F	T

The lack of an equivalence here (which mirrors an analogous lack of equivalence in the case of negated *conjunctions* encountered in § 2.6) is scarcely surprising. The first sentence says, "It's not the case that either Fenton or Iola will investigate"; the second, "Either Fenton won't investigate or Iola won't investigate." The first sentence denies that either will investigate; the second is consistent with one, though not both, investigating.

─── **Exercises 2.08** ───

Translate the following sentences into $\mathcal{L}_s$. Let C = Chet loses the map; F = Frank calls home; G = Gertrude finds the treasure; I = Iola finds the treasure; J = Joe brings a shovel.

1. Either Iola or Gertrude finds the treasure.

2. Neither Iola nor Gertrude finds the treasure.

3. If Chet doesn't lose the map and Joe brings a shovel, then Iola or Gertrude finds the treasure.

4. If neither Iola nor Gertrude finds the treasure, Frank calls home.

5. Chet doesn't lose the map, or Joe doesn't bring a shovel, or Frank doesn't call home.

6. It's not the case that both Iola and Gertrude find the treasure.

7. If either Chet loses the map or Joe doesn't bring a shovel, Iola doesn't find the treasure.

8. Iola or Gertrude finds the treasure if Chet doesn't lose the map.

9. Either Chet loses the map or Iola and Gertrude don't find the treasure.

10. Iola or Gertrude finds the treasure if Chet doesn't lose the map and Joe brings a shovel.

2.15 CONDITIONALS AND BICONDITIONALS

In introducing conditionals, we discovered a less-than-perfect fit between conditional surface structures in English and $\mathcal{L}_s$ conditionals. The appearance of a conditional "if...then..." phrase in an English sentence is not a reliable sign that the sentence actually expresses a conditional. This feature of English makes the translation of conditionals particularly challenging. Moreover, the *order* of sentences in a conditional construction makes a difference in a way it does not in conjunctions or disjunctions. Thus, it is easy to become confused about what counts as the antecedent of a given conditional and what counts as its consequent. Just as in any translation, you must somehow learn to *see through* the superficial structure of the target sentence to its deep structure.

As we discovered in § 2.08, English conditionals often occur in "reverse order": consequents first, then antecedents. Instead of saying

> If you put it into the oven, then it'll rise

we might just as well have said

> It'll rise if you put it into the oven.

Both sentences are translated

$$O \supset R$$

In translating such sentences, the trick is to locate the if-clause, and treat this as the antecedent. This does not mean that every occurrence of "if" signals the antecedent of a conditional. Only-ifs and ordinary "plain" ifs make very different logical contributions to the sentences in which they occur. The sentence

> It'll rise only if you put it into the oven

is a straightforward "...only if..." conditional translated

$$R \supset O$$

Remember: when confusion threatens, you can always go back for another look at the truth conditions of the target sentence. In the present case, that might include asking whether the "only-if" sentence would be obviously false if R were true and O false. If so, then it expresses a conditional.

Although conditionals are associated with the phrases "...only if..." and "if...then...," these are at most signs that a conditional *might* be present, not guarantees that one *is* present. Consider the sentence

If you wash the dishes, then I'll give you a dollar.

This sentence contains a conditional component—reflected in the fact that you will expect to be given a dollar if you wash the dishes. What you probably do not expect, however, is to be given a dollar even when you do *not* wash the dishes. In treating the sentence as a straightforward conditional, we allow it to be true under those very circumstances

W D	$W \supset D$
T T	T
T F	F
F T	T
F F	T

The sentence is counted true in the third row of the truth table, the row in which I give you a dollar even though you do not wash the dishes. You might *hope* that I would do that, but the sentence by itself gives you no grounds for that hope. This suggests that the English original, despite its conditional appearance, may in fact express a biconditional

I'll give you a dollar *if and only if* you wash the dishes.

As the truth table for the $\mathcal{L}_s$ version of this biconditional makes plain, you can expect to receive a dollar if you wash the dishes, but not otherwise

W D	$W \equiv D$
T T	T
T F	F
F T	F
F F	T

Sentences that *appear* to be conditionals in English sometimes express biconditionals instead. This is one reason for the impression that conditionals in $\mathcal{L}_s$ differ radically from those in English. True, $\mathcal{L}_s$ conditionals do not always serve to translate English sentences containing "...only if..." and "if...then..." constructions. This could be due to a quirk of English: *conditional surface structures are not reliably paired with conditional deep structures.*

Having noted these difficulties, you will be relieved to learn that none of this is going to matter very much. The character of L_s is such that we can afford to allow ourselves to "mistake" unobvious English biconditionals for conditionals. The reason for this becomes clear in the next chapter. For the present, recall that biconditionals can be thought of as *containing* conditionals. We introduced biconditionals by means of back-to-back— that is, conjoined—conditional pairs. A mistaken translation of a biconditional into a conditional in L_s does not miss the mark completely.

This does not mean that differences between conditionals and biconditionals can simply be ignored. Here is a safe policy: translate English sentences into L_s as biconditionals *only when they are obviously biconditionals*, only when they include some such phrase as "…if and only if…" or "…just in case…." The policy allows us to translate

If you wash the dishes, then I'll give you a dollar

into L_s as a garden-variety conditional

$$W \supset D$$

We reserve the biconditional for sentences in which it is explicit

You'll win if and only if you play hard.
A set is empty just in case it lacks members.

These sentences could be translated into L_s as

$$W \equiv P$$
$$E \equiv L$$

In reserving biconditional constructions in L_s for obvious biconditionals, we follow a time-honored logical tradition. This makes perfect sense given the uses to which a formal language like L_s is typically put. Bear in mind that we can always make our translations more subtle should the need arise.

Exercises 2.09

Translate the following sentences into L_s. Let C = Callie spots the thief; F = Fenton makes an arrest; G = Gertrude falls asleep; J = Joe wrecks the roadster.

1. If Callie spots the thief, Fenton makes an arrest.

2. Joe doesn't wreck the roadster if Gertrude doesn't fall asleep.

3. If Gertrude doesn't fall asleep and Joe doesn't wreck the roadster, then Fenton makes an arrest.

4. Joe wrecks the roadster only if Gertrude falls asleep.

5. If Joe wrecks the roadster, Gertrude falls asleep.

6. Only if Gertrude falls asleep does Joe wreck the roadster.

7. Joe wrecks the roadster if and only if Gertrude falls asleep.

8. Fenton makes an arrest just in case Callie spots the thief and Joe doesn't wreck the roadster.

9. If either Joe wrecks the roadster or Callie doesn't spot the thief, Fenton doesn't make an arrest.

10. If Joe wrecks the Roadster then Fenton makes an arrest if and only if Gertrude doesn't fall asleep.

2.16 TROUBLESOME ENGLISH CONSTRUCTIONS

Some common English turns of phrase can cause special problems when we set out to translate them into L_s. As always, a successful translation depends on our making clear to ourselves the truth conditions of the sentence to be translated. Errors occur when we fail to take the time to think through the meaning of a particular sentence.

Consider the sentence

> You'll gain weight unless you exercise.

The sentence says

> If you don't exercise, then you'll gain weight.

This sentence can be translated into L_s as

$$\neg E \supset W$$

The translation illustrates a feature of "unless" constructions generally. Typically, such constructions follow the pattern

> p unless q
> $\neg q \supset p$

The pattern is useful to remember. We are liable to be tempted to represent a sentence like the one above *incorrectly* as

> If you exercise, then you won't gain weight

expressible in L_s as

$$E \supset \neg W$$

This sentence might be thought to be implied by anyone asserting the original sentence, but it differs dramatically from that sentence. The easiest way to see that this is so is to look at a parallel case

> You'll be caught unless you run.

This sentence tells us

> If you don't run, then you'll be caught.

Following the pattern above, it translates into L_s as

$$\neg R \supset C$$

It does *not* tell us, however, that

If you are caught, then you don't run: $C \supset \neg R$

The original sentence leaves open the possibility that you are caught despite running; you might run into a trap, for instance, or be outrun, or fall down. A truth table comparison of the sentences shows they possess distinct truth conditions:

R C	$\neg R$	$\neg R \supset C$	$C \supset \neg R$
T T	F	T	F
T F	F	T	T
F T	T	T	T
F F	T	F	T

This truth table reveals something important about "unless" constructions. Examine the fourth column, which presents the truth conditions of the correct translation of the original sentence: $\neg R \supset C$. Compare the truth conditions shown in this column with those for the disjunction, $C \vee R$, shown below:

R C	$\neg R$	$\neg R \supset C$	$C \vee R$
T T	F	T	T
T F	F	T	T
F T	T	T	T
F F	T	F	F

The columns match, so the truth conditions of $\neg R \supset C$ and $C \vee R$ are the same. So what? Well, we are now in a position to simplify the principle we formulated earlier governing translation of "unless" constructions

p unless *q*

$p \vee q$

To apply this principle, first locate the "unless" clause, and rewrite the sentence in "pidgin-$\mathcal{L}_s$." Given the sentence

Unless it's stirred, the sauce will go bad

we might use S to represent "The sauce is stirred," and B to stand for "The sauce will go bad." The sentence, in pidgin-$\mathcal{L}_s$, is

$$B \text{ unless } S$$

which, following the principle, becomes

$$B \vee S$$

Intermediate pidgin-$\mathcal{L}_s$ representations are often useful in translating otherwise confusing sentences.

A different kind of problem crops up in sentences like the following:

> If Fenton investigates, then, even though
> Callie doesn't, Homer will be caught.

Here, a conditional sentence

> If Fenton investigates, then Homer will be caught

incorporates an embedded sentence

> Callie doesn't investigate.

The conditional seems clear enough, but what are we to do with this embedded sentence? Notice, first, that Callie's not investigating is not a *component* of the conditional, despite popping up in the midst of it. Suppose we translated the sentence

$$(F \wedge \neg C) \supset H$$

This translation makes Homer's being caught conditional on Fenton's investigating *and* Callie's not investigating. In the English original, Homer's being caught is conditional solely on Fenton's investigating

$$F \supset H$$

That Callie doesn't investigate is a piece of information added to, but not made a part of, the conditional. We can translate the whole sentence simply by adding $\neg C$ to the original conditional

$$(F \supset H) \wedge \neg C$$

This translation makes it clear that Homer's being caught is not conditional on Callie's not investigating.

In general, we can adopt this procedure when we encounter "although," "even though," and "despite the fact" constructions. Such constructions typically signal the insertion of a piece of information that is best treated as a separate conjunct. Thus, the sentence

> If Frank or Joe is kidnapped, then, despite
> the fact that Iola escapes, Chet will not be happy

should be translated as

$$((F \vee J) \supset \neg C) \wedge I$$

Although such translations will prove occasionally awkward, they are the best we can hope for in $\mathcal{L}_s$.

Exercises 2.10

Translate the following sentences into $\mathcal{L}_s$. Let C = Chet weighs anchor; F = Frank pilots *The Sleuth*; G = Gertrude is rescued; I = Iola signals; J = Joe wears a disguise.

1. Gertrude is rescued if Joe wears a disguise.

2. Iola signals unless Gertrude is rescued.

3. Gertrude isn't rescued unless Iola signals.

4. Unless Chet weighs anchor, Frank doesn't pilot *The Sleuth*.

5. Joe wears a disguise unless either Iola signals or Gertrude is rescued.

6. If Iola signals, then Joe wears a disguise unless Gertrude is rescued.

7. Joe wears a disguise although Iola doesn't signal.

8. Iola signals but Gertrude isn't rescued even though Joe wears a disguise.

9. If Frank pilots *The Sleuth*, then, even though Iola doesn't signal, Gertrude is rescued.

10. Frank pilots *The Sleuth*, despite the fact that Joe wears a disguise, if Chet weighs anchor.

2.17 TRUTH TABLE ANALYSES OF $\mathcal{L}_s$ SENTENCES

Truth tables were introduced in the course of characterizing truth functional connectives in $\mathcal{L}_s$. We can summarize those definitions in a single grand table:

p q	$\neg p$	$p \wedge q$	$p \vee q$	$p \supset q$	$p \equiv q$
T T	F	T	T	T	T
T F	F	F	T	F	F
F T	T	F	T	T	F
F F	T	F	F	T	T

Given these characterizations, we can specify the truth conditions for any $\mathcal{L}_s$ sentence. This, as we noted earlier, is a consequence of the fact that $\mathcal{L}_s$ is a truth functional language: the truth conditions of every sentence are determined by the truth conditions of its constituent sentences.

Knowing the truth conditions for a sentence amounts to knowing its truth value in all possible worlds, where the rows in a truth table represent mutually exclusive and exhaustive sets of possible worlds. Such knowledge is less impressive than it sounds. We all know the truth conditions of the sentence

The winning number in next week's State Lottery is 566,792.

We know this sentence is true in every world in which 566,792 is selected as the winning ticket, and false in all other worlds. Were we to formulate a sentence of this sort mentioning every ticket for the State Lottery, we would know the truth conditions for each of these sentences as well. What we would not know is which ticket will win, which of the myriad possible worlds we effortlessly consider is the *actual* world.

Picture God deliberating about *which* world to create, and settling on one of these—the *best possible* world, of course. In fixing the truth value of every sentence, God thereby fixes the world. We might imagine God scrutinizing an immense truth table, one with countless rows, each row representing a possible world. In choosing a world, God chooses one of these rows and, simultaneously, the truth value of every sentence.

We cannot pretend to approximate this level of knowledge, although we are often in a position to narrow down the possibilities. We can say, then, with some justification, that our world belongs to a certain *set* of possible worlds, a set that might itself be infinite. This may be all we need to settle the truth values of sentences. The idea is nicely illustrated in $\mathcal{L}_s$. Given a sentence, together with a specification of its truth conditions, we can determine its truth value provided we know the truth values of its atomic constituents. Suppose we encounter the sentence

It's not the case that either Iola and
Callie escape or that Fenton investigates.

Translated into $\mathcal{L}_s$ it becomes

$$\neg((I \wedge C) \vee F)$$

Suppose, further, that we know that C and F are true, and that I is false. This information, together with what we know about the truth functional connectives $\neg$, $\wedge$, and $\vee$, enables us to calculate the truth value of the sentence with ease

I	C	F	$I \wedge C$	$(I \wedge C) \vee F$	$\neg(I \wedge C) \vee F$
T	T	T	T	T	F
T	T	F	T	T	F
T	F	T	F	T	F
T	F	F	F	F	T
F	T	T	F	T	F
F	T	F	F	F	T
F	F	T	F	T	F
F	F	F	F	F	T

Since we know that C and F are true, and I false, we know that the actual world belongs to the set of possible worlds picked out by the fifth row of the truth table. On that row, the sentence turns out to be *false*. The method is simple and, providing we are careful, foolproof. It is, however, unforgivably *tedious*. For sentences well-stocked with distinct atomic constituents, it would be decidedly more tedious. Recall that if a sentence contains n distinct atomic sentences, a truth table depicting its truth conditions will have 2^n rows. A sentence containing six distinct atomic sentences, for instance, would require a 64-row truth table.

Happily, we can make use of a much simpler method of calculating the truth values of complex $\mathcal{L}_s$ sentences given that we know the truth values of each of their atomic constituents. We construct what amounts to a single row of the relevant truth table, the row that incorporates the truth values of the atomic constituents of the target sentence.

Consider the sentence

$$\neg((J \wedge \neg F) \supset (C \wedge J))$$

Suppose we know that F and J are true, and that C is false. We then:

1. Write out the sentence and enter that sentence's truth value directly under each atomic sentence.

$$\neg((J \wedge \neg F) \supset (C \wedge J))$$
$$\quad\quad \text{T} \quad\ \text{T} \quad\ \text{F} \ \text{T}$$

2. Chart the values of every negated atomic sentence.

$$\neg((J \wedge \neg F) \supset (C \wedge J))$$
$$\quad\quad \text{T} \quad\ \text{T} \quad\ \text{F} \ \text{T}$$
$$\quad\quad\quad\quad\ \text{F}$$

3. Working "inside-out," chart the values of molecular sentences.

$$\neg((J \wedge \neg F) \supset (C \wedge J))$$
$$\quad\quad \text{T} \quad\ \text{T} \quad\ \text{F} \ \text{T}$$
$$\quad\quad\quad\quad\ \text{F}$$
$$\quad\quad\quad \text{F} \quad\quad\quad \text{F}$$

4. You will eventually arrive at a determination for the sentence as a whole.

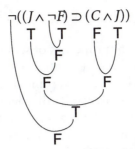

In this example, the truth value of the sentence is finally settled by the negation sign that includes the remainder of the sentence within its scope. Recall that the scope of a negation sign includes just the sentence to its immediate right. A negation sign to the left of an atomic sentence negates only that atomic sentence. A negation sign to the left of a pair of parentheses negates the sentence included within those parentheses.

Exercises 2.11

Determine the truth values of the L_s sentences below. Assume that A, B, and C are true; and that P, Q, R, and S are false.

1. $(Q \wedge B) \vee (\neg R \vee S)$ 2. $(\neg Q \wedge (B \vee S)) \supset P$

3. $((P \wedge Q) \supset \neg A) \supset S$ 4. $A \supset ((B \wedge \neg C) \vee (P \vee Q))$

5. $\neg(\neg(\neg P \wedge A) \supset ((B \vee Q) \wedge R))$

6. $((A \wedge B) \wedge P) \equiv (B \supset (C \vee S))$

7. $\neg((A \supset (P \vee C)) \wedge \neg(\neg B \vee \neg Q))$

8. $P \supset ((A \vee \neg B) \wedge (B \supset (C \vee Q)))$

9. $(S \vee (\neg Q \wedge \neg(R \vee S))) \equiv (\neg A \vee \neg(B \supset (\neg R \vee P)))$

10. $\neg((B \vee (P \wedge \neg Q)) \supset ((A \wedge \neg B) \vee (\neg P \supset (Q \vee R))))$

2.18 CONTRADICTIONS AND LOGICAL TRUTHS

Truth and falsity are *semantic* properties of sentences. Semantic properties are to be distinguished from a sentence's *syntactic* or grammatical properties. Every language has a syntax, principles specifying what does and what does not count as a sentence in that language. Most languages—certainly all natural languages—have a set of semantic principles that link the language to its subject matter as well. When we learn English, we learn the meanings of particular words. In the case of common nouns, for instance, we learn that "tree" is used to designate trees, that "star" is used to designate stars, and that "foot" is used to designate feet. But we also learn how words can be combined in sentences and used to make statements about actual and possible states of affairs. This requires our catching on to semantic principles governing sentences.

L_s, we have discovered, does not contain words—although you might think of it as employing one-word sentences at the atomic level. Sentences are built from sentences by means of truth functional connectives. A grasp of the semantics of L_s requires an understanding of the effects these connectives have on the truth conditions of sentences in which they figure. Just as in English we learn that the sentence

Snow is white and grass is green

is true just in case "Snow is white" is true and "Grass is green" is true, so we learn that in L_s

$$A \wedge B$$

is true if and only if A is true and B is true. The truth functional character of L_s allows us to specify these combinatorial principles simply and economically by means of truth tables. Principles governing English and other natural languages are more complicated. Although we have all somehow mastered them, no one has yet succeeded in spelling them out in detail.

We have seen that it is possible to know the truth *conditions* for a sentence without knowing its truth *value*, without knowing *whether* the sentence is true or false. In most cases, knowing a sentence's truth value requires that we know something about the world. If we know that the sentence

Snow is white

is true, that is because we know something about *snow*: its color in the actual world. Not all sentences are like this. Some sentences are such that if we know their truth conditions, we thereby know their truth values. Consider the sentence

Snow is white or snow isn't white.

Anyone who understands this sentence knows that it could not *fail* to be true. Its truth value depends not on the way the world is, but on the sentence's *syntax*. Given our usual understanding of conjunction and negation, we know that the sentence must be true. The point is clear if we translate the sentence into L_s

$$W \vee \neg W$$

and construct a truth table for it

W	$\neg W$	$W \vee \neg W$
T	F	T
F	T	T

The sentence is true no matter what truth values we assign to W, its lone atomic constituent; it is true in all possible worlds. A sentence true in all possible worlds is true in the actual world. Sentences that are true in all possible worlds are called *logical truths*. Logical truths expressible in L_s are called *tautologies*, although not all logical truths are tautologies.

Sometimes, as in the case above, the tautologous character of a sentence is obvious. This is not always the case, however. Consider the English sentence

If Desdemona loves Cassio, then, if the Moon is made
of Cheshire cheese, Desdemona loves Cassio

translated into L_s as

$$D \supset (M \supset D)$$

This sentence, too, is a logical truth, although that may not be obvious until we construct a truth table

D M	$M \supset D$	$D \supset (M \supset D)$
T T	T	T
T F	T	T
F T	F	T
F F	T	T

The table reveals what intuition might not: the sentence is true regardless of the truth values of D and M; it is true in all possible worlds.

Tautologous sentences differ, on the one hand, from ordinary *contingent* sentences and, on the other hand, from *contradictions*. A contingent sentence in L_s is a sentence whose truth value varies with assignments of truth values to its atomic constituents. More generally, a contingent sentence is true in some possible worlds and false in some possible worlds. Most of the L_s sentences we have encountered thus far have been contingent. Had we constructed truth tables for each of them, we would have discovered that these sentences were true in at least one row of their respective truth tables and false in at least one row. The truth table for a contradictory sentence, in contrast, shows that it is *false* at every row, and consequently that such sentences are false in all possible worlds.

Contradictions may be explicit, as in

Snow is white and snow is not white

or unexplicit, as in

Desdemona loves Cassio and it's not the case that either the Moon
isn't made of Cheshire cheese or Desdemona loves Cassio.

If we translate these sentences into L_s and construct an appropriate truth table for each, we can see that each is contradictory, each is false on every row of its truth table. The truth table for the first sentence looks like this:

W	$\neg W$	$W \wedge \neg W$
T	F	F
F	T	F

The second sentence's truth table looks like this:

D M	$\neg M$	$\neg M \vee D$	$\neg(\neg M \vee D)$	$D \wedge \neg(\neg M \vee D)$
T T	F	T	F	F
T F	T	T	F	F
F T	F	F	T	F
F F	T	T	F	F

The technique in $\mathcal{L}_s$ is simple and mechanical. For any sentence, we can decide whether it is contingent, tautologous, or contradictory simply by constructing a truth table.

In the case of English sentences, matters are often less obvious. No comparable mechanical technique for working out the truth conditions of English sentences is available. As a result, we sometimes find that sentences we have supposed were contingent are in fact contradictory or logically true. A contradictory sentence cannot be true, so it cannot be used to impart information about the world. Similarly, logically true sentences fail to impart information by failing to distinguish among possible worlds. If I tell you that snow is white, I provide you with information about the color of snow. You know now that our world is a member of the set of possible worlds in which snow is white. If I say "Snow is white or snow is not white," what I have said is true but informationally empty. You cannot tell, on the basis of what I have said, *anything* about the character of the world.

If our interest is in imparting information, it is not enough that what we say is true. Interesting truths are those that *might have been false.* A scientific hypothesis that is not "falsifiable," one that is not false in some possible worlds, is uninformative. The informativeness of a hypothesis is proportional to the ratio of worlds in which it is false to those in which it is true. A *maximally informative* hypothesis is one true in just one possible world. Logically true hypotheses are maximally *un*informative. It is sometimes claimed that psychoanalytic theories have this character. Suppose a diagnosis is pronounced correct when it elicits the patient's assent (the assent proving it true) *and* correct when it produces a denial (by being "defensive" the patient has proved the diagnosis correct). What could make it false? Hypotheses that *could not* be false are informationally empty.

Exercises 2.12

Determine which of the $\mathcal{L}_s$ sentences that follow are tautologous (logically true), which are contradictory, and which are contingent by constructing a truth table for each.

1. $P \supset (\neg P \supset P)$

2. $(A \wedge \neg A) \supset Q$

3. $A \vee (B \supset (A \wedge C))$

4. $\neg(P \supset (\neg P \supset P))$

5. $(R \vee (P \wedge Q)) \equiv \neg(\neg R \wedge \neg(P \wedge Q))$

6. $((P \supset Q) \wedge (Q \supset R)) \supset (P \supset R)$

7. $((P \wedge \neg(Q \vee R)) \supset (A \wedge (P \supset S)) \wedge (S \supset (P \vee (Q \wedge R)))$

8. $((B \vee A) \wedge \neg B) \wedge \neg A$

9. $(A \supset (B \supset C)) \wedge (B \supset (A \supset \neg C))$

10. $\neg(T \supset \neg(S \wedge \neg P)) \supset ((T \wedge S) \vee P)$

2.19 DESCRIBING $\mathcal{L}_s$

An obvious method of offering a description of something is to provide an exhaustive inventory of its constituent parts together with an account of the relations these parts bear to one another. I might describe the chair in which I am sitting, for instance, by listing its parts—arms, legs, seat, and back—and indicating how each of these parts is related to the rest. The technique is workable only when the object of a description is neither excessively large nor excessively complex. What is required for the description of sprawling abstract objects like English or $\mathcal{L}_s$? Both English and $\mathcal{L}_s$ comprise an infinite number of sentences, so we have no prospect of listing them all.

We might think of a description as a recipe that would enable us to decide, for any object whatever, whether it was or was not the thing described. A description of Paris distinguishes Paris from every other city; a description of Socrates picks Socrates out of the set of philosophers. In the same way, a description of English might incorporate a method for distinguishing English sentences from everything else. In the case of L_s, we can provide such a distinguishing recipe that is both rigorous and elegant. The recipe is founded on a set of rules specifying the *syntax* of L_s and thereby enabling us to determine, for any object at all, whether it is or is not a sentence of L_s. Such rules might be thought of as *generating* all the sentences of L_s and only these: using finite means, they give birth to an infinite object.

The syntactic rules of L_s provide us with a *recursive definition* of "sentence of L_s." Once formulated, we can make use of these rules in providing an account of the *semantic* features of L_s. We might think of the syntax and semantics of L_s as *components* of our description of L_s. The idea is perfectly general. We might set out to describe English or any other natural language by providing a generative syntactic component and a semantic component. To be sure, natural languages are more intricate than L_s. Linguists and logicians have managed to provide syntactic and semantic analyses for narrowly circumscribed portions of English, but much remains uncharted. In contrast to English, L_s is *completely* charted.

2.20 THE SYNTAX OF L_s

In specifying the syntax of L_s, we begin with an account of the vocabulary. Here we resort to the technique of listing elements. Although L_s contains an infinite number of sentences, these are constructed from a finite, hence listable, vocabulary. Let us call the set of vocabulary elements **V**. **V** comprises three sets of elements:

> **S:** sentential constants, $\{A, B, C, ..., Z\}$;
> **C:** truth functional connectives, $\{\neg, \wedge, \vee, \supset, \equiv\}$;
> **P:** left and right parentheses, $\{(,)\}$.

You can gain a sense of the task ahead by thinking of the ways in which the elements of **V** might be combined into finite strings. Imagine that all of these finite sequences are contained in a set, **V***. If we allow for the repetitions of elements, **V*** will have infinitely many members, among them:

A
$\neg C$
$(P \supset Q) \vee M$
$)(G \neg \equiv$
$\supset \supset)Z$

Plainly, some of the members of V^* are $\mathcal{L}_s$ sentences, and some are not. Our task is to provide a way of specifying, for any member of V^*, whether it is or is not a sentence of $\mathcal{L}_s$. Think of V^* as an infinte set of strings of symbols that includes a set of strings, V^*, the sentences of $\mathcal{L}_s$. The relation of V to V^* is illustrated below.

Note that although $\mathcal{L}_s$ is a proper subset of V^* (that is, V^* includes V and includes, as well, strings not in V), it too is infinite.

It will simplify our characterization of the syntax of $\mathcal{L}_s$ if we first define a technical notion that will enable us to keep the formulation of syntactic rules uncomplicated. Let us say that a string of symbols enclosed between matching left and right parentheses is *bounded*. Examples of bounded strings include:

$(A \supset B)$
$(G \neg \vee)$
$(\neg(\equiv)$

Each of these strings is a member of V^*, although only the first qualifies for membership in $\mathcal{L}_s$. This is obvious to anyone who has used $\mathcal{L}_s$, but our task is to make explicit the intuitions on which we rely in making such judgments.

The rules that follow specify the requirements for sentencehood in $\mathcal{L}_s$. All and only those members of V^* that satisfy these rules are sentences of $\mathcal{L}_s$, and hence members of V.

1. Every member of S is a sentence of $\mathcal{L}_s$;
2. If p is a sentence of $\mathcal{L}_s$, then $\neg p$ is a sentence of $\mathcal{L}_s$;
3. If p and q are sentences of $\mathcal{L}_s$, then $(p \wedge q)$ is a sentence of $\mathcal{L}_s$;
4. If p and q are sentences of $\mathcal{L}_s$, then $(p \vee q)$ is a sentence of $\mathcal{L}_s$;

5. If p and q are sentences of $\mathcal{L}_s$, then $(p \supset q)$ is a sentence of $\mathcal{L}_s$;
6. If p and q are sentences of $\mathcal{L}_s$, then $(p \equiv q)$ is a sentence of $\mathcal{L}_s$;
7. Nothing is a sentence of $\mathcal{L}_s$ that is not constructed in accord with rules 1-6.

Recursive Definition

The class of sentences of $\mathcal{L}_s$ is infinite. Even so, we can uniquely specify that class by means of a *recursive definition*.

A recursive definition proceeds by first characterizing an initial member (or finite collection of initial members) of the class. Second, rules are given for generating the remaining members of the class from these initial members.

A recursive rule is framed in a way that permits it to apply to its own output. If we begin with something stipulated to be a member of the class and apply an appropriate recursive rule to it, a new member of the class is generated. If we apply the rule to this generated member, the result is another member of the class—and so on.

The rules characterize an infinite object, **V**, the set of sentences in $\mathcal{L}_s$. To see how they work, let us apply them to a simple sentence

$$(\neg(G \wedge L) \supset (\neg H \equiv K))$$

By rule 1, we know that G, L, H, and K are all sentences of $\mathcal{L}_s$: all belong to S, the set of sentential constants. Rule 2 tells us that $\neg H$ is a sentence of $\mathcal{L}_s$—a negated sentence is a sentence. The strings $(G \wedge L)$ and $(\neg H \equiv K)$ count as sentences of $\mathcal{L}_s$ by rules 3 and 6, respectively. $(G \wedge L)$ is a sentence of $\mathcal{L}_s$, so $\neg(G \wedge L)$ is an $\mathcal{L}_s$ sentence, again by rule 2. Finally, given that $\neg(G \wedge L)$ and $(\neg H \equiv K)$ are sentences of $\mathcal{L}_s$, $(\neg(G \wedge L) \supset (\neg H \equiv K))$ is an $\mathcal{L}_s$ sentence by rule 5.

The first six rules tell what objects count as sentences of L_s. Rule 7 tells us that *only* these objects count as sentences of L_s. Rule 7 is crucial for drawing *boundaries* around the set of L_s sentences, *excluding* members of V^* that ought to be excluded. Taken together, the seven rules *define* "sentence of L_s": they are satisfied by, and only by, sentences of L_s.

The *recursive* character of our rules is illustrated by rule 2. Given just the first two rules, we can generate an infinite set of L_s sentences. By rule 1 we know that F, for instance, is a sentence of L_s. Rule 2 tells us that a negated sentence is a sentence, hence that $\neg F$ is a sentence of L_s. Since $\neg F$ is a sentence, rule 2 tells us that $\neg\neg F$ is a sentence; because *that* is a sentence, $\neg\neg\neg F$ is a sentence, too, *and so on*. We are entitled to the "...and so on" here, since the rule equips us to generate endless sentences by means of the simple device of piling up—concatenating—negation signs.

Given our definition, most of the expressions we have called sentences of L_s in this chapter are not in fact sentences. Rules 3 through 6 require sentences containing truth functional connectives (other than the negation sign) to be *bounded*, that is, to be enclosed within parentheses. The expression

$$R \supset B$$

does not count as a sentence. We could *turn it into* a sentence by adding parentheses

$$(R \supset B)$$

We can respond to this turn of events in one of three ways. First, we could bite the bullet and insist that alleged sentences be converted to genuine sentences by the addition of appropriate parentheses. Second, we could rewrite our rules, complicating them slightly, so as to allow unbounded strings to count as sentences of L_s. Third, we could adopt the informal convention of omitting "needless" parentheses. Let us adopt— retroactively—this third option.

2.21 THE SEMANTICS OF L_s

The semantics of a formal language like L_s revolves around the notion of an *interpretation*. The syntactic rules of L_s provide us with an infinite collection of *uninterpreted* strings of symbols. In using L_s, we assign meanings to sentential constants, members of the set S, and we make use of truth table characterizations of connectives to extend meanings to complex sentences.

Our initial assignment of meanings to the elements of **S** amounts to an assignment of truth values to those elements. An interpretation of $\mathcal{L}_s$ is (1) an assignment of truth values to sentential constants, and (2) the use of truth table characterizations of connectives to produce truth values for every sentence. Why is there a connection between interpretation and truth? Reflect on what happens when we decide to let

W

represent

Socrates is wise.

In so interpreting *W*, we in effect assign it a truth value. *W* will be true if and only if Socrates *is* wise. Had we interpreted *W* differently, *W* might well have had a *different* truth value.

With this background, we can proceed to define the notion of an interpretation for $\mathcal{L}_s$:

> An interpretation, *I*, of $\mathcal{L}_s$ is an assignment to each member of **S** (that is, to each atomic sentence) one and only one of the truth values {*true, false*} and an assignment to the members of **C** (that is, to the connectives) the truth functional meanings set out below.

An interpretation might be thought of as a single row of a truth table in which every atomic sentence is assigned a truth value. Given *n* atomic sentences, the truth table would have 2^n rows representing 2^n possible combinations of truth values of atomic constituents. In characterizing $\mathcal{L}_s$, we have restricted its atomic vocabulary to the uppercase letters, *A* through *Z*. Thus characterized, a truth table representation of an interpretation of $\mathcal{L}_s$ would consist of one of 2^{26} or 67,108,864 rows of a truth table roughly the height of the Washington Monument. We might have permitted the vocabulary of $\mathcal{L}_s$ to be expanded, perhaps by adding numerical subscripts to alphabetic characters. Were this done, the number of truth value combinations could become literally uncountable. We shall ignore this possibility.

We are now in a position to define *true under an interpretation* for $\mathcal{L}_s$. Where *I* is an interpretation of $\mathcal{L}_s$, and *p* and *q* are sentences:

> 1. If *p* is a member of **S**, then *p* is true under *I* if and only if *I* assigns the value *true* to *p*.

2. $\neg p$ is true under I if and only if p is not true under I.

3. $(p \wedge q)$ is true under I if and only if p is true under I and q is true under I.

4. $(p \vee q)$ is true under I if and only if p is true under I or q is true under I, or both.

5. $(p \supset q)$ is true under I if and only if q is true under I or p is not true under I, or both.

6. $(p \equiv q)$ is true under I if and only if either p and q are both true under I or neither p nor q is true under I.

7. p is false under I if and only if p is not true under I.

Except in rule 1, p and q need not be atomic sentences.

Truth

Pilate asked, "What is truth?" Philosophers have provided a number of answers in the form of definitions or "theories" of truth. In general, our grasp of the notion of truth is better than our grasp of these theories.

The logician, Alfred Tarski, provided an important definition of truth in "The Concept of Truth in Formalized Languages" [reprinted in many places, including *The Logic of Grammar*, Donald Davidson and Gilbert Harmon, eds., (Encino, CA: Dickenson Publishing Co., 1974)]. Tarski's account of truth allows for a rigorous specification of the semantics of formal languages.

For our purposes, truth can be treated as a primitive, undefined notion, one that can be used to define other notions.

Given this definition of truth under an interpretation, we can extend the semantics of $\mathcal{L}_s$. First, we define the notion of a *model*:

An interpretation, I, is a *model* of a sentence, p, if and only if $I(p)$ is true. If Γ (*gamma*) is a set of sentences, then I is a model of Γ if and only if every sentence in Γ is true under I.

An interpretation is a function from sentences to truth values, so the value of $I(p)$ is the truth value an interpretation I assigns to a sentence p. If I is a model of a sentence, p, (or a set of sentences, Γ) then I assigns the value *true* to p (or to every member of Γ).

The notion of a model enables us to provide straightforward definitions of other semantic terms:

> A sentence, p, is *logically true* if and only if every interpretation is a model of p (that is, p is true under every interpretation).

> A sentence, p, is *contradictory* if and only if p has no model (that is, p is false under every interpretation).

> A sentence, p, is *contingent* if and only if there is at least one model of p and at least one interpretation that is not a model of p (that is, p is true under some interpretations, and false under others).

The concept of an interpretation and the related concept of a model will prove useful in subsequent chapters. For the present, let us bear in mind that every language possesses a syntax—a principled distinction between sentences and meaningless strings of elements of the language—and a semantics—a function associating meanings or truth conditions and sentences. Mastery of a language requires mastery of both. It requires something more, as well. It is to this *something more* that we now turn.

3

Derivations in $\mathcal{L}_s$

3.00 SENTENTIAL SEQUENCES

Our understanding of sentences in $\mathcal{L}_s$ has only just begun. This may come as something of a shock, but consider: we credit others with understanding a word only if they can use that word appropriately in sentences. Sentences, like words, bear relations to one another. We recognize that the sentence

If Socrates is wise, then he is happy

and the sentence

Socrates is wise

stand in a special relation to one another and to the sentence

Socrates is happy.

Anyone who failed to appreciate this relationship—anyone who did not see that the third sentence *follows from* the others—could not be said to have mastered those sentences.

This chapter is devoted to an investigation of the logical relations sentences bear to one another. The focus will be not on individual sentences but on *sequences* of sentences. Some sequences—*derivations*—exhibit a particular structure. These structures are governed by rules just as sentences themselves are governed by rules. Learning to use these rules requires learning how to construct derivations in $\mathcal{L}_s$, and thus finally to master the language.

3.01 OBJECT LANGUAGE AND METALANGUAGE

Whenever we discuss a language, we must *use* some language or other. This situation sometimes gives rise to ambiguities and confusions. Philosophers are often accused of arguing about "mere words." Whether this is true or not, philosophers have unquestionably tended to succumb to a special class of confusions that appear purely linguistic. St. Augustine relates the following conversation:

> *A*: He was a terrible man. Manure came out of his mouth.
> *B*: Is that possible?
> *A*: Well, he said "Manure smells bad," and whatever he said must have come out of his mouth. So manure came out of his mouth.

Some theorists have spent years on problems that, because they were based on a confusion of the sort illustrated here, were specious.

Quickly, then, let us distinguish between the use of language to talk *about language* and the use of language to talk about nonlinguistic reality. In the conversation recounted by Augustine, *A* conflates these uses. That is, *A* runs together talk about a *word*, "manure," and talk about that bit of non-linguistic reality designated by the word "manure," *manure*. *A* is guilty of a *use–mention* confusion. The distinction between use and mention is an instance of a general distinction that comes into play whenever we pause to consider *representations*. In discussing cases in which some item—a word, a symbol, a picture, a thought—functions representationally, we are obliged to distinguish between representations or symbols themselves and what they represent or symbolize. When we begin representing representations, the sort of confusion illustrated by St. Augustine can easily arise.

Confining ourselves to languages like English or $\mathcal{L}_s$, we can distinguish between an *object language*—a language under discussion—and a *metalanguage*—a language in which the discussion of some object language is conducted. The distinction is relative: no language is, in its own right, an object language or a metalanguage. A language may be either, depending on whether it is being *talked about*—in which case it is the object language—or being *used* to talk about another language—in which case it is the metalanguage. If we discuss $\mathcal{L}_s$ in English then, for purposes of that discussion, $\mathcal{L}_s$ is the object language and English the metalanguage. Were we to formulate truths about English in $\mathcal{L}_s$, these roles would be reversed: $\mathcal{L}_s$ would be the metalanguage and English the object language. Potential for confusion arises not so much in cases in which object language and metalanguage are clearly distinct, but in those cases in which a language is used to talk about itself, where one language is *both* metalanguage and object language.

When this occurs, we can minimize confusion by adopting conventions that signal our intentions. When the aim is to discuss, in English, a particular piece of English, that piece might be marked in some way—by the use of quotation marks, for instance, or by placing it on a line by itself. If I want to talk about the sentence "Socrates is wise," I may put quotation marks around it, as I have done here, or put it on a separate line

Socrates is wise.

In so doing, I make it clear that my aim is to focus on this linguistic item and not on the state of affairs that the linguistic item represents.

Logicians sometimes say that by putting an expression within quotation marks, we turn the expression into the *name of itself.* Consider, for instance, the sentence

The postal employee has 17 letters.

Compare it to the sentence

"The postal employee" has 17 letters.

The first sentence says something about the postal employee, the second about *the expression* "the postal employee." These sentences pretty obviously have different truth conditions, as do the following:

Bol is shorter than Muggsie.
"Bol" is shorter than "Muggsie."

By deploying quotation marks judiciously and by setting expressions off, we can avoid many use–mention confusions. Unfortunately, we cannot avoid them all. Consider the sentence

This sentence is false.

What are the truth conditions for such a sentence? The sentence is, or certainly seems to be, paradoxical: if false, it is true; if true, it is false. Quotation marks are no help here. The fact that English, and for that matter any other natural language, includes such sentences has been taken as a sign that natural languages are themselves intrinsically flawed. We can reserve judgment on this difficult question and note that L_s is not similarly criticizable. The paradoxical sentence above cannot be translated into L_s.

The White Knight's Song

In *Through the Looking Glass*, Lewis Carroll depicts a conversation between Alice and the White Knight:

"You are sad," the Knight said in an anxious tone: "let me sing you a song to comfort you."

"Is it very long?" Alice asked, for she had heard a good deal of poetry that day.

"It's long," said the Knight, "but it's very, *very* beautiful. Everybody that hears me sing it—either it brings the *tears* into their eyes, or else—"

"Or else what?" said Alice, for the Knight had made a sudden pause.

"Or else it doesn't, you know. The name of the song is called 'Haddocks' Eyes'."

"Oh, that's the name of the song, is it?" Alice said, trying to feel interested.

"No, you don't understand," the Knight said, looking a little vexed. "That's what the name is *called*. The name really *is* '*The Aged Aged Man*'."

"Then I ought to have said "That's what the *song* is called"?" Alice corrected herself.

"No, you oughtn't: that's quite another thing! The *song* is called "*Ways and Means*": but that's only what it's *called*, you know!"

"Well, what *is* the song, then?" said Alice, who was by this time completely bewildered.

"I was coming to that," the Knight said. "The song really *is* '*A-sitting On A Gate*': and the tune's my own invention."

Explain the White Knight's distinctions. Is he entitled to them?

[See Peter Heath, *The Philosopher's Alice* (New York: St. Martin's, 1974), p. 218.]

Exercises 3.00

Add quotation marks to the sentences below in such a way that, if it is possible, (1) they remain grammatical and (2) they express plausible truths. (Thanks to John Corcoran.)

1. An acorn contains corn.

2. One plus one is not identical with two.

3. Without a cat there would be no catastrophe.

4. Sincerity involves sin.

5. French is not French.

6. One is not identical with one.

7. George said I am not a wimp.

8. Sentence 9 is true.

9. Sentence 8 is false.

10. I love the sound of a cellar door.

3.02 DERIVATIONS IN $\mathcal{L}_s$: FIRST STEPS

Derivations in $\mathcal{L}_s$ consist of sequences of sentences satisfying the following definition:

> A derivation is an ordered set of sentences, $\langle P,c \rangle$, such that (1) $\langle P,c \rangle$ is finite; (2) every member of P is either a premise or is introduced by means of an established rule; (3) c consists of a single sentence.

The definition tells us that a derivation is a finite ordered sequence of sentences consisting of *premises*, sentences themselves *derived from* premises, and a *conclusion* consisting of a single sentence. The angle brackets, ⟨ and ⟩, enclose items that occur in a particular order: premises precede the conclusion.

In constructing a derivation, we make explicit the *connection* between premises and conclusion; we show that the conclusion *logically follows from* the premises. In so doing, we provide a *proof* for the validity of the sequence. For this reason, derivations are often called *proofs*. We can use the terms interchangeably, although, as we shall see, proofs of validity need not take the form of derivations.

In English, a derivation might take the following form:

> If Socrates is wise, then he is happy.
> Socrates is wise.
> Therefore, Socrates is happy.

The sentences above the horizontal line are premises. The sentence below the line is a conclusion. If the argument is a good one, the conclusion is a *logical consequence* of the premises—alternatively, the premises *logically imply* the conclusion.

Sentences are evaluated with respect to their truth or falsity. Arguments are not true or false but *valid* or *invalid.* These notions are best characterized in stages. First, we define the notion of logical implication (and introduce a new symbol):

> A derivation, ⟨*P,c*⟩, is valid if and only if *P* logically implies *c* ($P \vDash c$).

The symbol $\vDash$ (lazy *pi*) stands for "logically implies." $P \vDash c$ means that P logically implies *c*. To complete the characterization of validity, we define the notion of logical implication:

> A set of sentences, *P*, logically implies a sentence, *c*, ($P \vDash c$) if and only if *c* cannot be false if *P* (i.e., every sentence in *P*) is true.

When *P* logically implies *c*, then *c* is said to be a *logical consequence* of *P*.

The notion of logical implication and the related notion of validity have a central place in the present chapter. Let us reflect on these notions before venturing further.

When a sentence or set of sentences logically implies a sentence, the relation is the relation we have in mind when we recognize that a sentence *follows from* another sentence or set of sentences. We can define this relation by referring to the truth conditions of sentences, their truth values across possible worlds. We can think of q following from or being logically implied by p when in no possible world is p true and q false. If p and q make up a sequence in which p is a premise and q is the conclusion, then, when p logically implies q, the sequence is valid. These definitions do *not* require that the premises or the conclusion of a valid sequence be *true*. They require only that *if* the premises are true, the conclusion is true. If the conclusion of a valid sequence is false, one or more of the premises on which it rests must be false as well. Validity is *truth preserving*, but not *truth guaranteeing*: if we begin with true premises and validly derive a conclusion from those premises, then the conclusion, too, must be true. If we know only that a conclusion follows validly from a set of premises, however, we do not thereby know whether the conclusion is true.

One way to prove that a sequence in L_s is valid is to construct a derivation. Before discussing the mechanics of derivations, however, let us look at another technique for proving validity, one that relies solely on truth tables. Consider the following sequence:

> If human beings are fish, then whales are fish.
> <u>Human beings are fish.</u>
> Therefore, whales are fish.

The reasoning in this sequence is perfectly valid. The conclusion, "Whales are fish," follows from the premises: in no possible world are the premises true and the conclusion false. Yet the conclusion and, in this case, one of the premises are false. Suppose we translate the relevant sentences into L_s

$$H \supset W$$
$$\underline{H \qquad}$$
$$W$$

Once in L_s, we can construct a truth table to prove validity

H	W	H ⊃ W
T	T	T
T	F	F
F	T	T
F	F	T

The truth table contains no row in which the premises, H and $H \supset W$, are true and the conclusion, W, is false. In two rows, the second and fourth, the conclusion is false, but in those rows one of the premises is false as well.

Consider another English sequence

> If Socrates drinks the hemlock, then he dies.
> <u>Socrates doesn't drink the hemlock.</u>
> Therefore, Socrates doesn't die.

Is the reasoning in this sequence valid? Does the conclusion follow from—is it implied by—the premises? Is there a possible world in which the premises are true and the conclusion false? Our intuitions may waver. Once we put the sequence into $\mathcal{L}_s$, however, the answer becomes clear

H	D	$H \supset D$	$\neg H$	$\neg D$
T	T	T	F	F
T	F	F	F	T
F	T	T	T	F
F	F	T	T	T

The third row reveals a possible world in which the premises of the original sequence are true and its conclusion false. That world might be one in which hemlock is lethal (as it is in the actual world) and in which Socrates does not drink the hemlock but dies anyway—from some other cause.

An argument of this sort in fact commits the well-known fallacy of *denying the antecedent*. From a conditional sentence, together with the denial of its antecedent, we cannot infer the denial of the consequent. In everyday discussion we sometimes fail to pick up fallacious—that is, *invalid*—arguments because we are distracted by our knowledge of the truth or falsity of their premises and conclusions. Yet it is perfectly possible for all of the premises *and* the conclusion of an argument to be true and for the argument still to be invalid. The last row of the truth table above illustrates precisely this possibility. It is important, then, in evaluating arguments in English, to attend to the logical relation holding between premises and conclusion and learn to ignore the isolated truth values of their constituent sentences.

Before leaving the topic of logical implication, let us take a moment to distinguish this notion from ordinary conditionals. You may be tempted to read sentences of the form

$$p \supset q$$

as "p implies q." If "implies" is taken to mean "logically implies," this reading is incorrect. Consider a simple conditional sentence in English

If the sky is blue, then the Earth is round.

This sentence concerns the color of the sky and the shape of the Earth. It is, moreover, true—in the actual world its antecedent and consequent are both true. In contrast, the sentence

"The sky is blue" implies "the Earth is round"

concerns not the Earth and sky, but, as the quotation marks indicate, a pair of *sentences*. It is, in addition, *false*: The sentence

The sky is blue

does not imply—that is, logically imply—the sentence

The Earth is round.

You can easily envisage a possible world in which the former is true and the latter false.

Fallacious Reasoning

We reason fallaciously when we offer arguments in which premises fail to imply conclusions. In an effort to combat fallacious reasoning, logicians and rhetoriticians have compiled lists of common mistakes or fallacies. The fallacy of denying the antecedent is one of these, as is the complementary fallacy of affirming the consequent, illustrated by the following argument:

If it's raining, then the streets are wet.
The streets are wet.
Therefore, it's raining.

You can prove that this argument is invalid by translating it into L_s, constructing a truth table, and locating a row in that truth table in which the premises are true and the conclusion is false.

Exercises 3.01

For each of the $\mathcal{L}_s$ sequences below, construct a truth table that demonstrates its validity or invalidity, one that shows whether its premises do or do not logically imply its conclusion. If the sequence is invalid, circle the row or rows that establish this fact.

1. $P \supset Q$
   $R \supset P$
   R
   ___
   Q

2. $P \supset Q$
   $P \supset R$
   ___
   $Q \supset R$

3. $P \supset Q$
   $\neg(Q \wedge R)$
   ___
   $P \supset \neg R$

4. $P \vee \neg Q$
   $\neg(Q \wedge \neg R) \supset \neg S$
   $\neg P$
   ___
   $\neg Q$

5. $P \wedge Q$
   $Q \supset (P \vee R)$
   ___
   R

6. $P \supset (Q \supset P)$
   ___
   P

7. $(P \wedge Q) \supset R$
   $P \wedge \neg R$
   ___
   $\neg Q$

8. $P \supset Q$
   $R \supset Q$
   R
   ___
   P

9. $P \supset Q$
   Q
   ___
   P

10. $P \vee Q$
    P
    ___
    $\neg Q$

11. $P \wedge Q$
    $P \supset (R \vee \neg Q)$
    ___
    R

12. $P \supset Q$
    $\neg Q$
    ___
    $\neg P$

13. $P \supset Q$
    $R \supset S$
    $P \vee R$
    ___
    $Q \vee S$

14. $P \supset Q$
    $\neg Q$
    ___
    $P \supset R$

15. $(P \vee Q) \supset R$
    $P \vee S$
    $\neg S$
    ___
    R

3.03 DERIVATIONS IN $\mathcal{L}_s$: THE PRINCIPLE OF FORM

Logic is silent as to the truth or plausibility of premises and conclusions. Its province is validity. A valid argument need not have a true conclusion. But if an argument incorporates true premises and if its conclusion follows validly from those premises, then that conclusion must be true. An ideal knower needs both true assumptions and valid inferential procedures.

Philosophers since Descartes (1596–1650) have dreamed of basing all knowledge on a few indisputable truths. Those truths, together with appropriate inferences, would yield a complete system of knowledge. Cartesian foundationalism is out of fashion, but the need for valid inferences is universal and timeless.

Algorithms

An algorithm is a principle that guarantees a particular result in a finite number of steps.

Think of an algorithm as prescribing a step-by-step procedure for achieving a previously specified goal. Each step is entirely mechanical, calling for no "insight" or intelligence. In this sense, algorithmic solutions to problems are "mindless," although considerable intelligence may have been required to invent them.

Imagine that you want to find your way through a maze. The maze is finite, and you are indifferent to the time it takes to locate the exit. You can guarantee success by following a simple algorithm: *turn left at every corner.* Although the procedure will lead you up blind alleys, it does not require that you remember which paths you have taken. You need only be able to identify corners, to execute the instruction to turn left, and to recognize the exit when you eventually reach it.

Truth tables afford algorithmic proofs of validity or invalidity in L_s. You can mechanically establish that a sequence is valid by constructing a truth table for the sequence and noting whether, on any row of the table, the premises are true and the conclusion is false. If no such row appears, the sequence is valid; it is invalid otherwise.

Truth tables provide a mechanical technique for proving the validity of sequences in L_s. As the exercises above clearly illustrate however, truth tables are unwieldy and tedious. Although they afford exhaustive analyses of

$\mathcal{L}_s$ sequences, truth tables lend themselves to notational errors. Worse, truth table proofs of validity are applicable only to sequences that can be smoothly translated into $\mathcal{L}_s$. Much, perhaps most, of the reasoning we use every day outstrips the meager resources of $\mathcal{L}_s$.

A more intelligent, interesting, and natural way to demonstrate validity is to construct a derivation. Derivations are founded on inferential principles that extend beyond $\mathcal{L}_s$. A derivation comprises a sequence of sentences, including premises and a conclusion, in which the reasoning from premises to conclusion has been made explicit. The derivational component of $\mathcal{L}_s$ has been designed to resemble familiar informal patterns of reasoning. $\mathcal{L}_s$ constitutes what logicians call a *natural deduction system*, one in which derivational proofs of validity are founded on "natural" patterns of reasoning.

Derivations in $\mathcal{L}_s$ depend on a *principle of form*:

> If a sequence, $\langle P,c \rangle$, is valid, then any sequence with the same form is valid.

The key to understanding the principle is in understanding what it means for two sentences, or sentence sequences, to have *the same form*. We can acquire an intuitive grip on this notion by considering examples.

Return to the valid sequence discussed earlier

> If human beings are fish, then whales are fish.
> Human beings are fish.
> Therefore, whales are fish.

The $\mathcal{L}_s$ representation of this sequence is

$$H \supset W$$
$$\underline{H }$$
$$W$$

The principle of form tells us that if this sequence is valid, all sequences with the *same form* as this one are valid. What, then, is the *form* of the sequence? Suppose we set out the sequence using *variables*

$$p \supset q$$
$$\underline{p }$$
$$q$$

The sequence consists of a pair of premises, the first of which is a conditional sentence and the second, the antecedent of that conditional; and a conclusion that consists of the consequent of the conditional. Expressed informally, this is the *form* of the sequence. The following sequences have the same form as the original sequences:

> If the sky is blue, then the Earth is round.
> The sky is blue.
> Therefore, the Earth is round.

In $\mathcal{L}_s$

$$B \supset R$$
$$\underline{B\qquad}$$
$$R$$

The sameness of pattern is obvious in these examples. That is not always the case. The $\mathcal{L}_s$ sequence below has the same form as those above:

$$\neg(A \vee (C \supset D)) \supset (D \wedge \neg C)$$
$$\underline{\neg(A \vee (C \supset D))}$$
$$(D \wedge \neg C)$$

This might surprise you. In one respect, this complex sequence looks *nothing at all* like the original sequence. Closer examination reveals that the sequence, like the earlier one, consists of (1) a conditional sentence, (2) the antecedent of that conditional, followed by (3) its consequent. The antecedents and consequents are themselves nonatomic, but insofar as we are concerned with the sequence's form, that is irrelevant. The sequence matches the pattern

$$p \supset q$$
$$\underline{p\qquad}$$
$$q$$

You will be in a position to understand derivations in $\mathcal{L}_s$ once you learn to *see* such patterns. The construction of derivations is a *perceptual* task far more than it is an intellectual one (see chapter one). It is akin to bird watching or the solving of picture puzzles. As in those endeavors, patterns that are invisible at first come to seem obvious with practice.

3.04 INFERENCE RULES: *MP, MT,* AND *HS*

We have seen that the sequence below is valid:

$$H \supset W$$
$$\underline{H}$$
$$W$$

By generalizing the pattern we obtained a generalized sequence of the form

$$p \supset q$$
$$\underline{p}$$
$$q$$

The principle of form tells us that every sequence with this form is valid. The pattern of inference exhibited in this sequence is both common and natural, so we can adopt it as a derivation rule, calling it by its traditional Latin name *modus ponendo ponens*—for short, *modus ponens,* or *MP.*

$$(MP) \quad p \supset q, p \vdash q$$

This formulation of *MP* indicates that $p \supset q$ and p *deductively yield q.* That is, *MP* permits us to derive q from $p \supset q$ and p, where p and q are understood to be variables ranging over any $\mathcal{L}_s$ sentences. This rule, together with other rules, will be used in the derivation of conclusions from premises. The mechanics of derivations are best learned by example, so let us work through a simple derivation, commenting on its features.

Suppose we set out to prove the sequence mentioned earlier

$$\neg(A \lor (C \supset D)) \supset (D \land \neg C)$$
$$\underline{\neg(A \lor (C \supset D))}$$
$$(D \land \neg C)$$

Modus ponens enables us to support the move from the two premises—the sentences above the horizontal line—to the conclusion. To facilitate the procedure, let us assign each line in the derivation a number. Numbers will function as labels for the lines on which they occur. When we need to refer to a particular sentence in a derivation, we can do so by means of the number of the line on which the sentence occurs.

Line numbers contribute to the intelligibility of derivations. Intelligibility is aided by the use of other devices as well. To the left of every premise, we place a plus sign, +, to signal that the sentence in question is a premise. To the left of the conclusion to be derived, we enter a question mark, ?, thereby indicating that the sentence is taken to follow deductively from the premises. In the course of constructing derivations, we often find it necessary to add sentences that have been derived by means of a rule from other sentences. To the right of every sentence so added we place (1) the name of the rule that supports its insertion, and (2) the line number or numbers of sentences to which the rule was applied. Putting this all together, we can construct a derivation for the preceding sequence as follows:

1. + $\neg(A \vee (C \supset D)) \supset (D \wedge \neg C)$
2. + $\neg(A \vee (C \supset D))$
3. ? $(D \wedge \neg C)$
4. $(D \wedge \neg C)$ 1, 2, MP

The sentences in lines 1 and 2 are premises, as the + to their left indicates; line 3 contains the conclusion we intend to derive. A conclusion, always signalled by a ?, is never itself used in a derivation. It is inserted following the premises as a way of indicating the *goal* of the derivation. The sentence on line 4 is derived; hence we must support its introduction with the name of the rule that entitles us to enter it, *MP*, together with the line numbers of sentences to which the rule applies. The notation in line 4 indicates that $(D \wedge \neg C)$ was derived from sentences in lines 1 and 2 via *MP*.

Derivation rules will be introduced in two stages. Rule *MP* is an example of an *inference rule*. Inference rules mirror principles of reasoning we all employ unself-consciously in our everyday thinking. This is less so in the case of *transformation rules*. In practice, transformation rules enable us to manipulate $\mathcal{L}_s$ sentences so as to achieve a pattern suitable for the application of one or more inference rules. Although the logical relationships exemplified by transformation rules are commonplace, the rules themselves may not seem to resemble natural patterns of reasoning. We might think of transformation rules as pattern synthesizing devices. They will be discussed below (§§ 3.09–3.13).

Rules are formulated, not in $\mathcal{L}_s$, but in the metalanguage. Rules tell us when it is permissible to derive particular sentences from other sentences. For this reason rules are formulated using *variables*: *p*s and *q*s in rules do not stand for atomic $\mathcal{L}_s$ sentences, but for any $\mathcal{L}_s$ sentence whatever. This feature is what enabled us to construct the derivation above.

We have seen how the rule *modus ponens* (*MP*) functions. This rule, like every other derivation rule, applies only to *whole lines* of derivations. The following application of *MP* violates this restriction:

1. $B \vee (S \supset W)$
2. S
3. W 1, 2, MP *(invalid)*

MP cannot be applied to *portions* of sentences. The reason should be clear: the conditional sentence is itself part of a disjunction. Suppose I tell you

> Either the sky is blue or, if I sell you a lottery
> ticket for $5, you will win the lottery.

You would be foolish to infer that were I to sell you a ticket, you would win the lottery. You would be foolish to infer that even if you had every reason to believe that my disjunctive utterance was truthful. The moral is that in $\mathcal{L}_s$, rules of implication apply only to whole sentences, not to sentences that are themselves parts of sentences.

Inference and Implication

Implication is a relation among sentences. Inference is a mental act. "If A then B," coupled with "A," logically implies "B." This means that if both "If A then B" and "A" are true, "B" must be true (there is no possible world in which the first two sentences are true and the third is not true).

Your accepting the first two sentences above does not thereby rationally oblige you to infer or accept the third. At most, reason demands your acceptance of "B" or your rejection of either "If A then B" or "A."

Ordinary sentences, singly and in combination, imply endless other sentences. We take the trouble to infer these other sentences only when we have some reason to do so.

A rule similar to *modus ponens, modus tollendo tollens*—for short, *modus tollens (MT)*—enables us to infer the denial of the antecedent of a conditional sentence given (1) the conditional sentence, and (2) the denial of its consequent

$$(MT) \quad p \supset q, \neg q \vdash \neg p$$

Consider a parallel sequence in English

> If Socrates is wise, then he is happy.
> Socrates isn't happy.
> Therefore, Socrates isn't wise.

The pattern of reasoning is straightforward and familiar. Consider its deployment in $\mathcal{L}_s$

$$W \supset H$$
$$\neg H$$
$$\neg W$$

We can construct a derivation of this sequence, marking each premise with a +, using a ? to indicate the conclusion to be derived, and showing that the conclusion does follow from the premises by *MT*

1. $+$ $W \supset H$
2. $+$ $\neg H$
3. $?$ $\neg W$
4. $\neg W$ 1, 2, *MT*

Derivations are not always so straightforward. Consider the English sequence

> If it's not a holiday, then the freeway is jammed.
> The freeway isn't jammed.
> Therefore, it's a holiday.

The sequence consists of a conditional sentence, "If it's not a holiday, then the freeway is jammed," followed by the denial of the consequent of that conditional, "The freeway isn't jammed." The conclusion, "it's a holiday," permitted in accord with *modus tollens*, is the denial of the antecedent. Why not "It's not the case that it isn't a holiday?" We recognize that negated negations are equivalent to unnegated sentences: "It's not the case that it isn't a holiday" is logically equivalent to "It's a holiday." This is easily proved via a truth table:

p	$\neg p$	$\neg\neg p$
T	F	T
F	T	F

The truth conditions of p are the same as those for $\neg\neg p$.

─ Derivation Heuristics ─

A *heuristic* is a rule of thumb, a principle that can prove useful in the solution of a problem, but that does not guarantee a solution. Heuristics are distinguished from algorithms, solution-guaranteeing procedures.

We have seen that truth tables serve as algorithms for the assessment of validity or invalidity in $\mathcal{L}_s$. By constructing a truth table, we can "mindlessly" establish that a sequence is valid or invalid.

Matters are different when we set out to prove validity by means of derivations. There is no "mindless" technique for applying rules of inference to premises in such a way as to guarantee the derivation of the conclusion of every valid sequence. Failure to derive a conclusion could mean either that the sequence is invalid or that, although it is valid, we have not yet stumbled on a sequence of rule applications that shows its validity.

This does not mean that the construction of derivations is a matter of blind luck. Once we have the knack, we can often see how a derivation might go. Acquiring the knack involves the acquisition of pattern recognition skills: we learn to see—in the literal sense—familiar patterns in the midst of unfamiliar arrays of symbols.

One rule of thumb or heuristic commonly thought to help in the construction of derivations is this:

- Focus on the conclusion and "work backwards" to the premises.

In so doing, you may narrow to a manageable few the range of options in the application of rules.

How might this feature of negation best be captured in $\mathcal{L}_s$? We could introduce a rule—*double negation (DN)*—that would allow us to infer p

from $\neg\neg p$. The derivation below reflects the application of such a rule, together with MT to an $\mathcal{L}_s$ version of the English sequence above:

$$
\begin{array}{lll}
\text{1.} & + \ \neg H \supset J & \\
\text{2.} & + \ \neg J & \\
\text{3.} & ? \ \ H & \\
\text{4.} & \ \ \ \neg\neg H & \text{1, 2, } MT \\
\text{5.} & \ \ \ H & \text{4, } DN \\
\end{array}
$$

Although DN could be included in the list of derivation rules for $\mathcal{L}_s$, we can avail ourselves of a less cumbersome mechanism that will streamline many derivations.

Every sentence has a *valence*, positive or negative. A negated sentence has a negative valence; a sentence not preceded by a negation sign has a positive valence. The sentences below have a positive valence:

$$
\begin{array}{c}
P \\
P \vee Q \\
P \supset (Q \vee \neg R)
\end{array}
$$

The valence of the following sentences is negative:

$$
\begin{array}{c}
\neg P \\
\neg (P \vee Q) \\
\neg (P \supset (Q \vee \neg R))
\end{array}
$$

Let us agree to interpret negation signs appearing in rules as instructions to *reverse the valence* of expressions to which they are attached. MT applies when we are given a conditional sentence together with the negation of its consequent—that is, together with a sentence identical with its consequent but with the opposite valence. From this we infer the negation of its antecedent—that is, we infer a sentence identical with the antecedent but with the opposite valence. Reversing the valence of a *negated* sentence yields an unnegated sentence. This means that in the derivation above, we can infer H *directly* from lines 1 and 2 via MT.

$$
\begin{array}{lll}
\text{1.} & + \ \neg H \supset J & \\
\text{2.} & + \ \neg J & \\
\text{3.} & ? \ \ H & \\
\text{4.} & \ \ \ H & \text{1, 2, } MT \\
\end{array}
$$

The same point applies to negated consequents, as illustrated by the derivation below:

$$
\begin{aligned}
&1. \;+\; A \supset \neg B \\
&2. \;+\; B \\
&3. \;? \;\; \neg A \\
&4. \qquad \neg A \qquad\qquad 1, 2, MT
\end{aligned}
$$

MT permits the derivation of the negation of the antecedent of a conditional, given the negation of its consequent. The consequent of the conditional $A \supset \neg B$ is $\neg B$. Reversing the valence of $\neg B$ yields B.

MP and MT mirror everyday forms of reasoning with conditional sentences. A third rule, *hypothetical syllogism* (HS), involving conditionals, governs inferences of the sort illustrated below:

> If Socrates is wise, then he is happy.
> <u>If Socrates is happy, then he is satisfied.</u>
> Therefore, if Socrates is wise, then he is satisfied.

The reasoning here has the form

$$
(HS) \quad p \supset q, q \supset r \vdash p \supset r
$$

Such sequences consist of a pair of conditionals in which the consequent of one matches the antecedent of the other, from which we are permitted to derive a third conditional. This characteristic of conditionals illustrates their *transitivity*: if r is conditional on q, and q is conditional on p, then r is conditional on p.

A derivation of the sequence above would look like this:

$$
\begin{aligned}
&1. \;+\; W \supset H \\
&2. \;+\; H \supset S \\
&3. \;? \;\; W \supset S \\
&4. \qquad W \supset S \qquad\qquad 1, 2, HS
\end{aligned}
$$

Rule HS applies whether the conditionals involved are simple or complex, and regardless of their order of occurrence in a derivation. The derivation below illustrates both possibilities:

$$
\begin{aligned}
&1. \;+\; \neg(A \vee \neg B) \supset (C \supset D) \\
&2. \;+\; (R \wedge S) \supset \neg(A \vee \neg B)
\end{aligned}
$$

3. ? $(R \wedge S) \supset (C \supset D)$

4. $(R \wedge S) \supset (C \supset D)$ 1, 2, *HS*

Here, as elsewhere, we must acquire the knack of seeing patterns amidst sometimes bewildering arrays of symbols.

---Exercises 3.02---

Construct derivations for the $\mathcal{L}_s$ sequences below using MP, MT, and HS, and interpreting occurrences of $\neg$ in MT as an instruction to reverse the valence of the expression to the right of the $\neg$.

1. $+ P \supset (Q \vee R)$
   $+ \neg (Q \vee R)$
   $? \neg P$

2. $+ \neg P \supset \neg Q$
   $+ \neg P$
   $? \neg Q$

3. $+ \neg P \supset Q$
   $+ \neg Q$
   $? P$

4. $+ P \supset Q$
   $+ \neg S$
   $+ \neg (Q \supset R) \supset S$
   $? P \supset R$

5. $+ P \supset \neg Q$
   $+ Q$
   $+ \neg S \supset P$
   $? S$

6. $+ P \supset (Q \supset R)$
   $+ P$
   $+ Q$
   $? R$

7. $+ \neg (P \supset Q) \supset R$
   $+ \neg R$
   $+ P$
   $? Q$

8. $+ P \supset Q$
   $+ Q \supset R$
   $+ P$
   $? R$

9. $+ P \supset \neg Q$
   $+ Q$
   $+ R \supset P$
   $? \neg R$

10. $+ \neg (P \supset R) \supset \neg Q$
    $+ P$
    $+ Q$
    $? R$

11. $+ \neg P \supset \neg Q$
    $+ Q$
    $+ P \supset (S \wedge T)$
    $? S \wedge T$

12. $+ P \supset Q$
    $+ Q \supset \neg R$
    $+ R$
    $? \neg P$

13. $+ \neg (P \supset Q) \supset R$
    $+ R \supset S$
    $+ \neg S$
    $? P \supset Q$

14. $+ \neg (P \wedge \neg S) \supset (Q \vee R)$
    $+ (Q \vee R) \supset \neg T$
    $+ T$
    $? P \wedge \neg S$

15. $+ (P \supset S) \supset T$
    $+ T \supset \neg (Q \vee \neg R)$
    $+ ((P \supset S) \supset \neg (Q \vee \neg R)) \supset U$
    $? U$

3.05 RULES FOR CONJUNCTION: $\wedge I$ AND $\wedge E$

We have looked at three derivation rules pertaining to conditionals. Now we turn to a pair of rules involving conjunction. The first rule, *conjunction introduction* ($\wedge I$), permits the derivation of a conjunction of two sentences given those sentences:

> Socrates is wise.
> Socrates is happy.
> Therefore, Socrates is wise and happy.

The sequence translated into in $\mathcal{L}_s$ incorporates an application of $\wedge I$

1. W
2. H
3. $W \wedge H$ 1, 2, $\wedge I$

We can express $\wedge I$ more generally as follows:

$$(\wedge I) \quad p, q \vdash p \wedge q$$

A second rule, *conjunction elimination* ($\wedge E$), goes in the reverse direction. From

> Socrates is wise and happy

we are permitted to infer each of the sentences

> Socrates is wise

and

> Socrates is happy.

The rule is formulated in two parts, making this option explicit:

$$(\wedge E) \quad p \wedge q \vdash p$$
$$p \wedge q \vdash q$$

Applying $\wedge E$ to an $\mathcal{L}_s$ version of the English sequence above enables us to "bring down" conjuncts individually.

1. $W \land H$
2. W 1, $\land E$
3. H 1, $\land E$

Conjunction rules, like all other derivation rules, apply only to whole lines in derivations. From

$$\neg(W \land H)$$

we cannot derive either a W or an H. Such an inference ignores the negation sign and is clearly invalid, as the English sequence below illustrates:

> <u>It's not the case that Socrates is wise and happy.</u>
> Therefore, Socrates is wise.

We could prove the invalidity of the L_s sequence by constructing a truth table. This would show that there are worlds in which the premise is true and the conclusion is false.

Exercises 3.03

Construct derivations for the L_s sequences below using MP, MT, HS, $\land$I, and $\land$E.

1. $+ (P \supset Q) \land \neg R$
   $+ P$
   $? Q$

2. $+ P \land (\neg Q \land \neg R)$
   $? \neg R$

3. $+ \neg(P \land \neg Q) \supset R$
   $+ \neg R$
   $? \neg Q$

4. $+ \neg P$
   $+ Q$
   $+ (\neg P \land Q) \supset R$
   $? R$

5. $+ P \land Q$
   $? Q \land P$

6. $+ P \supset (Q \land \neg R)$
   $+ P$
   $? \neg R$

7. $+ P \supset (Q \land \neg R)$
   $+ P$
   $? \neg R \land Q$

8. $+ (P \land Q) \supset (R \land S)$
   $+ Q$
   $+ P$
   $? R$

9. $+ \neg(P \wedge Q) \supset (P \vee Q)$
   $+ \neg(P \vee Q)$
   $? Q$

10. $+ S \wedge ((P \equiv Q) \supset R)$
    $+ P \equiv Q$
    $? R$

11. $+ P \supset Q$
    $+ R \supset S$
    $+ P \wedge R$
    $? Q \wedge S$

12. $+ P \supset Q$
    $+ Q \supset (R \wedge S)$
    $+ P \wedge T$
    $? S \wedge T$

13. $+ (P \wedge Q) \supset \neg(S \wedge T)$
    $+ \neg S \supset \neg T$
    $+ T$
    $? \neg(P \wedge Q)$

14. $+ P \supset (Q \supset \neg R)$
    $+ P \wedge Q$
    $? \neg R$

15. $+ S \wedge \neg Q$
    $+ \neg(R \wedge (S \supset T)) \supset Q$
    $? T$

3.06 RULES FOR DISJUNCTION: $\vee I$ AND $\vee E$

Disjunction rules differ from those affecting the $\wedge$ connective in a way that mirrors the truth functional difference between disjunction and conjunction. *Disjunction introduction* ($\vee I$) permits us to add any sentence we please to a given sentence. The added sentence need not be present earlier in the derivation; it need not be imported from some other line. The idea is straightforward. Suppose we know that

> Socrates is happy.

We can infer that

> Socrates is happy or wise

or, for that matter,

> Socrates is happy or whales are fish.

These disjunctions add no information to the sentences from which they were derived. They do not say, for instance, that, in addition to being happy,

Socrates is wise or that whales are fish. "Socrates is wise" and "Whales are fish" are present only as components of disjunctions. The rule, expressed in $\mathcal{L}_s$, has the form

$$(\vee I) \quad p \vdash p \vee q$$

In practice, we employ $\vee I$ when we need to add something not present already. If, in looking over a sequence, we notice that an atomic sentence occurs in the conclusion but is absent from the premises, the odds are good that $\vee I$ will come into play should we set out to construct a derivation. Rule $\vee I$ permits the addition of atomic sentences to atomic sentences

    1. H
    2. $H \vee W$ 1, $\vee I$

as well as the addition of nonatomic, molecular sentences

    1. H
    2. $H \vee \neg(W \supset \neg P)$ 1, $\vee I$

The rule permits the introduction of *any* sentence to a sentence already present, provided we honor the general restriction. Its application is restricted to whole lines, as the sequence below illustrates:

    1. $H \vee W$
    2. $(H \vee W) \vee \neg(W \supset \neg P)$ 1, $\vee I$

When $\neg(W \supset \neg P)$ is added to $H \vee W$, it is added to the *whole sentence*, not to W alone. The derivation below is defective:

    1. $H \vee W$
    2. $H \vee (W \vee P)$ 1, $\vee I$ (*incorrect*)

Disjunction introduction's companion rule, *disjunction elimination* ($\vee E$), differs from its conjunction counterpart, $\wedge E$, in requiring *two* sentences for its application. Given a disjunction, together with the denial of one of its disjuncts (that is, a sentence identical with one of its disjuncts but possessing the opposite valence), $\vee E$ permits us to derive the remaining disjunct. Thus, in English:

Either Socrates is wise or Socrates is happy.
<u>Socrates isn't wise.</u>
Therefore, Socrates is happy.

The pattern of reasoning invoked exhibits the following form:

$$W \vee H$$
$$\underline{\neg W}$$
$$H$$

The sequence below, which differs from the one above only in featuring the negation of the second disjunct, is equally sound:

Either Socrates is wise or Socrates is happy.
<u>Socrates isn't happy.</u>
Therefore, Socrates is wise.

For that reason, we formulate two versions of $\vee E$, just as we did for $\wedge E$.

$$(\vee E) \quad p \vee q, \neg p \vdash q$$
$$p \vee q, \neg q \vdash p$$

Applying $\vee E$ to the sequence above:

1. $+$ $W \vee H$
2. $+$ $\neg H$
3. ? W
4. $\quad W$ $\qquad\qquad$ 1, 2, $\vee E$

Again, we should read negation signs in the formulation of rules as pertaining to the *valences* of sentences, and not as denoting the occurrence of actual negation signs in sentences under consideration. The rule tells us that, given a disjunction, say

$$\neg R \vee \neg (S \wedge P)$$

together with a sentence identical with one of its disjuncts but with the opposite valence, say

$$R$$

we can infer the remaining disjunct

$$\neg(S \wedge P)$$

More formally

1. + $\neg R \vee \neg(S \wedge P)$
2. + R
3. ? $\neg(S \wedge P)$
4. $\neg(S \wedge P)$ 1, 2, $\vee E$

We cannot, however, apply the rule to a sentence that is itself contained within another sentence:

1. $P \wedge (Q \vee \neg R)$
2. R
3. $P \wedge Q$ 1, 2, $\vee E$ (*incorrect*)

This restriction, as we have seen, applies to every rule of inference.

Exercises 3.04

Construct derivations for the $\mathcal{L}_s$ sequences below using MP, MT, HS, $\wedge$I, $\wedge$E, $\vee$I, and $\vee$E.

1. + P
   + $(P \vee \neg R) \supset Q$
   ? Q

2. + $(P \vee Q) \supset (R \wedge S)$
   + P
   ? S

3. + $P \supset Q$
   + $P \vee \neg R$
   + R
   ? Q

4. + $P \supset (Q \vee R)$
   + $\neg(Q \vee R) \vee S$
   + $\neg S$
   ? $\neg P$

5. + $(P \wedge Q) \vee \neg(R \wedge S)$
   + S
   + R
   ? Q

6. + $P \supset \neg(Q \wedge R)$
   + $(Q \wedge R) \vee S$
   + $\neg S$
   ? $\neg P$

7. + $P \supset Q$
   + P
   ? $P \wedge (Q \vee R)$

8. + $P \supset (Q \vee \neg S)$
   + $P \wedge S$
   ? Q

9. $+ P$
    $+ \neg Q$
    $+ \neg (Q \lor R) \supset \neg P$
    $? R$

10. $+ P \supset (Q \land R)$
    $+ S \lor \neg T$
    $+ S \supset P$
    $+ T$
    $? Q$

11. $+ P \supset (Q \lor R)$
    $+ S \supset (P \land \neg R)$
    $+ S \land T$
    $? Q$

12. $+ P \supset Q$
    $+ (Q \lor (R \supset S)) \supset (S \lor T)$
    $+ \neg S \land P$
    $? T$

13. $+ (P \lor \neg Q) \supset R$
    $+ P$
    $? R \lor (S \equiv \neg T)$

14. $+ P \supset Q$
    $+ (Q \lor R) \supset (R \lor \neg S)$
    $+ P \land \neg R$
    $? \neg S$

15. $+ P \supset \neg (Q \lor R)$
    $+ Q \land S$
    $+ T \lor P$
    $? T$

3.07 CONDITIONAL PROOF (CP)

Imagine that you want to prove a conditional sentence, "If Socrates is a philosopher, then he is happy," and you have already established that all philosophers are happy. A natural way to reason is as follows:

> Suppose that Socrates is a philosopher; then it would follow that because all philosophers are happy, Socrates is happy; therefore, if Socrates is a philosopher, then he is happy.

In so reasoning, you derive a conditional sentence by (1) supposing its antecedent true "for the sake of argument," and (2) showing that, given your other premise, if this antecedent is true, the consequent is true. This amounts to establishing that the conditional sentence follows from the premises with which we began.

The pattern of reasoning is reflected in $\mathcal{L}_s$ by a *conditional proof rule* (*CP*). The rule allows us to enter a sentence, p, as a supposition. If q can be shown to follow from this supposition, then we are entitled to affirm the conditional sentence, $p \supset q$. The rule is set out as follows:

$$(CP) \quad \begin{bmatrix} p \\ \vdots \\ q \end{bmatrix}$$
$$p \supset q$$

The $\ulcorner$ marks a *suppositional premise*. It functions in derivations just as "suppose" functions in English sequences. The $\llcorner$ indicates that a supposition has been *discharged*.

Conditional proofs are useful in the derivation of conditional sentences. Conditionals can occur as conclusions or as sentences derived en route to the derivation of other sentences. *CP* permits us to introduce, *at any point in a derivation*, the antecedent of a conditional we intend to derive. The suppositional sentence, *p*, can be used in the derivation just as though it were a premise. The suppositional character of *p* is signalled by the presence of a $\ulcorner$. In deriving *q*, the consequent of the pertinent conditional, we discharge this supposition and "close off" the application of *CP* by placing a $\llcorner$ beside *q*. We then enter the conditional, $p \supset q$, the conditional, that is, whose antecedent consists of the sentence introduced as a supposition and whose consequent consists of the last sentence derived, the sentence to the right of the $\llcorner$. To indicate the scope of our supposition, we connect the $\ulcorner$ and $\llcorner$ as shown in the example below:

1. $+ \; S \supset M$
2. $? \; S \supset (S \wedge M)$
3. $\quad \begin{bmatrix} S \end{bmatrix}$
4. $\quad \; ? \; S \wedge M$
5. $\quad \; M$ 1, 3, *MP*
6. $\quad \; S \wedge M$ 3, 4, $\wedge I$
7. $\quad S \supset (S \wedge M)$ 3–6, *CP*

Line 3 introduces *S* as a supposition. A subgoal, $S \wedge M$, is introduced in line 4 to mark the sentence we intend to derive with the help of our supposition. Once this subgoal is derived, the supposition is discharged and the conditional proved on its basis is set out in line 7.

This derivation typifies the role of conditional proof in derivations. Again, *CP* can be used to derive *any* conditional sentence sought in a derivation. That sentence may, but need not, be the conclusion. Once a supposition is entered, attention shifts from the original conclusion to the subgoal, the consequent of the conditional we intend eventually to derive. When that consequent is established, however, and the supposition is bracketed—signalling that it has been discharged—the sentences contained within those brackets become *inactive*: they cannot be used in subsequent lines.

CP is a powerful and versatile rule. This becomes evident when we embed applications of CP inside one another. Embedded conditional proofs are useful for deriving conditional sentences that themselves contain conditional consequents. Imagine setting out to derive the sentence

$$(S \supset P) \supset (S \supset R)$$

We might enter as a supposition

$$S \supset P$$

and seek, as a subgoal the sentence

$$S \supset R$$

Since this subgoal is *itself* a conditional, we might establish it using an embedded CP strategy. The derivation below illustrates the technique:

1.	+ $P \supset (Q \vee R)$	
2.	+ $(S \wedge P) \supset \neg Q$	
3.	? $(S \supset P) \supset (S \supset R)$	
4.	⌈ $S \supset P$	
5.	| ? $S \supset R$	
6.	| ⌈ S	
7.	| | ? R	
8.	| | P	4, 6, MP
9.	| | $Q \vee R$	1, 8, MP
10.	| | $S \wedge P$	6, 8, $\wedge I$
11.	| | $\neg Q$	2, 10, MP
12.	| ⌊ R	9, 11, $\vee E$
13.	⌊ $S \supset R$	6–12, CP
14.	$(S \supset P) \supset (S \supset R)$	4–13, CP

Such derivations are at first daunting. With practice, they become second nature. We need only recognize that when a conditional sentence is needed in a derivation, we can often obtain the sentence by first introducing its antecedent as a supposition and then deriving its consequent. The strategy is applied in the derivation below to obtain a sentence that is then used to derive the conclusion:

1.	$+ P \supset (\neg Q \vee R)$	
2.	$+ Q$	
3.	$+ (P \supset R) \supset S$	
4.	$? S$	
5.	$\quad P$	
6.	$\quad ? R$	
7.	$\quad \neg Q \vee R$	1, 5, MP
8.	$\quad R$	2, 7, $\vee$E
9.	$P \supset R$	5–8, CP
10.	S	3, 9, MP

Here, rule *CP* is used to derive, not the conclusion of the sequence, *S*, but a sentence used to derive the conclusion, $P \supset R$.

Exercises 3.05

Construct derivations for the L_s sequences below using rules discussed thus far, including CP.

1. $+ P \supset Q$
   $+ P \supset (Q \supset R)$
   $+ Q \supset (R \supset S)$
   $? P \supset S$

2. $+ P \supset (S \supset (Q \wedge R))$
   $+ (Q \wedge R) \supset \neg P$
   $+ T \supset S$
   $? P \supset \neg T$

3. $+ P \supset Q$
   $+ R \supset S$
   $? (P \wedge R) \supset (Q \wedge S)$

4. $+ (P \wedge Q) \supset R$
   $+ P$
   $? Q \supset R$

5. $+ P \supset ((Q \vee R) \supset S)$
   $+ (S \vee T) \supset W$
   $? P \supset (Q \supset W)$

6. $+ P \supset (Q \vee R)$
   $+ P \supset \neg Q$
   $? P \supset R$

7. $+ (P \supset R) \supset (Q \supset S)$
   $+ Q$
   $? (P \supset R) \supset S$

8. $+ P \supset S$
   $+ R \supset S$
   $? P \supset (R \supset S)$

9. $+ (P \wedge R) \vee \neg S$
   $+ Q \vee T$
   $+ P \supset \neg Q$
   $? (S \supset T) \vee R$

10. $+ Q \supset (T \vee S)$
   $+ \neg R \wedge \neg T$
   $+ P$
   $? P \wedge (Q \supset S)$

11. $+ \neg P \lor (S \supset Q)$
    $+ S \land T$
    $? P \supset (Q \lor R)$

12. $+ (P \lor \neg T) \supset ((S \lor T) \supset Q)$
    $+ \neg P \lor S$
    $? (P \supset Q) \lor (S \supset T)$

13. $+ T \supset \neg P$
    $+ T \lor (S \lor R)$
    $+ \neg R$
    $? P \supset S$

14. $+ P \supset (S \lor T)$
    $+ (S \lor T) \supset (Q \supset (R \lor \neg S))$
    $+ S$
    $? P \supset (Q \supset R)$

15. $+ (Q \supset R) \supset \neg S$
    $+ T \supset (\neg (Q \supset R) \supset P)$
    $? S \supset (T \supset (P \lor Q))$

3.08 INDIRECT PROOF (*IP*)

A common strategy for establishing the truth of a claim is first to suppose that the claim is false, then show that this leads to an absurdity, and finally conclude that the claim must be true. Reasoning of this sort is widespread in ordinary life and in the sciences. Holmes argues that the suspect could not have been in London at the time of the murder by showing that the supposition that he *was* in London leads to an impossibility: had the suspect been in London at the time he could not have witnessed the accident in Berkshire. A mathematical proof that no largest prime number exists begins with the supposition that there *is* a largest prime and shows that this has paradoxical consequences.

This pattern of reasoning—proving a sentence true by supposing it false and showing that this supposition leads to an absurdity—is called *reductio ad absurdum* ("reduction to absurdity") or, more commonly in logic and mathematics, *indirect proof* (*IP*). In $\mathcal{L}_s$, the format of indirect proofs will resemble that used for *CP*

$$(IP) \quad \begin{array}{|l} \neg p \\ \vdots \\ q \land \neg q \end{array}$$
$$ p$$

Imagine that we want to prove p. Using *IP*, we introduce p's negation, $\neg p$, as a supposition, and then show that this supposition leads to a *paradox*:

$q \wedge \neg q$. A paradox, in this context, is an explicit contradiction, the conjunction of any sentence and the negation of that sentence. We thus discharge the supposition and enter the sentence we set out to prove. The sentences below are contradictory and hence, on this characterization, paradoxical:

$$P \wedge \neg P$$
$$(P \vee \neg Q) \wedge \neg (P \vee \neg Q)$$

In introducing a supposition, *IP*, like *CP*, introduces a subgoal. For *CP*, this subgoal is a sentence making up the consequent of a conditional sentence we intend to derive. In deploying *IP*, the subgoal is the derivation of some sentence—*any* sentence—together with its negation The indefinite character of the subgoal is signalled by an ×. Consider an application of *IP*:

1.	+ $P \vee R$	
2.	+ $R \supset S$	
3.	+ $\neg S \vee \neg R$	
4.	? P	
5.	$\neg P$	
6.	? ×	
7.	R	1, 3, $\vee E$
8.	S	2, 7, MP
9.	$\neg S$	3, 7, $\vee E$
10.	$S \wedge \neg S$	8, 9, $\wedge I$
11.	P	5–10, IP

We can derive S and $\neg S$ as shown in lines 8 and 9, respectively. These sentences contradict one another. In line 10, the contradiction is made explicit and the suppositional premise discharged. Bear in mind that the contradictory sentence can be *any* L_s sentence together with its negation, including the sentence we hope to derive—as in the example below:

1.	+ $P \supset (Q \vee R)$	
2.	+ $\neg P \supset R$	
3.	+ $\neg Q$	
4.	? R	
5.	$\neg R$	
6.	? ×	
7.	P	2, 5, MT
8.	$Q \vee R$	1, 7, MP
9.	R	3, 8, $\vee E$

10.	$R \wedge \neg R$	5, 9, $\wedge I$
11.	P	5–10, IP

IP and CP can be used together and embedded as the derivation below illustrates.

1. +	$\neg Q \vee (P \supset (R \vee S))$	
2. +	$\neg S \wedge \neg R$	
3. ?	$P \supset \neg Q$	
4.	P	
5.	$? \neg Q$	
6.	Q	
7.	$? \times$	
8.	$P \supset (R \vee S)$	1, 6, $\vee E$
9.	$R \vee S$	4, 8, MP
10.	$\neg S$	2, $\wedge E$
11.	R	9, 10, $\vee E$
12.	$\neg R$	2, $\wedge E$
13.	$R \wedge \neg R$	11, 12, $\wedge I$
14.	$\neg Q$	6–13, IP
15.	$P \supset \neg Q$	4–13, CP

Here, the derivation of a conditional sentence is accomplished by (1) entering the antecedent of the conditional as a supposition, and then (2) showing that the negation of the conditional's consequent leads to a contradiction. Note line 6 where Q is taken to be the negation of $\neg Q$. We need not enter $\neg\neg Q$ as the supposition, and then appeal to a principle of double negation, converting $\neg\neg Q$ to Q. It is less awkward to treat negation signs in applications of IP as we do in applications of other rules, as indicators of reversed valence. We should read IP as permitting the supposition of a sentence whose valence—its being negative or positive, negated or not—is the reverse of the sentence we intend to derive.

─── Exercises 3.06 ───

Construct derivations for the $\mathcal{L}_s$ sequences below using rules discussed thus far, including IP.

1. $+ P \supset Q$
   $+ R \supset P$
   $+ R \vee (Q \wedge S)$
   $? Q$

2. $+ P \vee Q$
   $+ P \supset (R \wedge S)$
   $+ (R \vee S) \supset Q$
   $? Q$

3. $+ (P \vee Q) \supset R$
   $+ \neg R$
   $? \neg P$

4. $+ \neg R \supset \neg(\neg P \vee Q)$
   $+ \neg R$
   $? P$

5. $+ (P \vee Q) \supset (R \supset \neg S)$
   $+ (S \vee T) \supset (P \wedge R)$
   $? \neg S$

6. $+ P \supset Q$
   $+ S \supset T$
   $? (P \vee S) \supset \neg(\neg Q \wedge \neg T)$

7. $+ P \supset ((Q \wedge R) \vee S)$
   $+ (Q \wedge R) \supset \neg P$
   $+ S \supset (T \supset \neg P)$
   $? P \supset \neg T$

8. $+ (P \vee Q) \supset (R \supset S)$
   $+ \neg P \supset T$
   $+ R \wedge \neg S$
   $? T$

9. $+ P \supset Q$
   $+ (Q \vee R) \supset S$
   $+ \neg S$
   $? \neg(P \vee S)$

10. $+ P \supset (\neg Q \wedge R)$
    $+ S \vee \neg T$
    $+ P \vee T$
    $? Q \supset S$

11. $+ S \supset (Q \vee R)$
    $+ S \vee (P \supset S)$
    $? P \supset (Q \vee R)$

12. $+ \neg(S \wedge T) \supset (Q \supset R)$
    $+ P \supset \neg T$
    $? P \supset (Q \supset R)$

13. $+ P \vee Q$
    $+ (P \vee R) \supset (S \supset T)$
    $+ S \wedge (T \supset Q)$
    $? Q$

14. $+ P \vee S$
    $+ S \supset (R \supset T)$
    $+ R \wedge (T \supset P)$
    $? P \vee Q$

15. $+ P \supset S$
    $+ P \vee (S \wedge T)$
    $? S \vee \neg T$

┌─────────── A Greatest Prime? ───────────┐

A prime number is a number divisible only by itself and 1. Is there a greatest prime, a prime number greater than any other prime number?

The Greek mathematician, Euclid (c. 300 B.C.) offered a famous *reductio* argument—indirect proof—that there is no greatest prime number:

1. Suppose there were a greatest prime, x.

2. Let y be the product of all primes less than or equal to x plus 1; y = (2 × 3 × 5 × 7 × ... × x) + 1.

3. If y is prime, then x is not the greatest prime because y is greater than x.

4. If y is not prime, y has a prime divisor, z, and z is different from each of the primes 2, 3, 5, 7,..., x less than or equal to x. Thus, z must be a prime greater than x.

5. But y is either prime or not prime.

6. Therefore, x is not the greatest prime.

7. Therefore, there is no greatest prime.

[See E. Nagel and J. R. Newman, *Gödel's Proof* (New York: New York University Press, 1967), pp. 37–38.]

└──┘

3.09 TRANSFORMATION RULES: *COM, ASSOC,* AND *TAUT*

The rules discussed thus far, rules of inference, enable us to derive sentences from sentences in the construction of derivations. Are these rules all we need in order to show, for any valid sequence in $\mathcal{L}_s$, that it is a valid sequence?

Consider the sequence below:

1. $+$ $P \supset Q$
2. $+$ $(R \vee Q) \supset S$
3. $+$ $\neg S$
4. ? $\neg(P \vee S)$

Suppose we set out to derive the conclusion using *IP*

1. $+$ $P \supset Q$
2. $+$ $(R \vee Q) \supset S$
3. $+$ $\neg S$
4. ? $\neg(P \vee S)$
5. $\lceil P \vee S$
6. $| \; ? \times$
7. $| \; P$ 3, 5, $\vee E$
8. $| \; Q$ 1, 7, *MP*
9. $| \; Q \vee R$ 8, $\vee I$

We can see that S, together with $\neg S$ in line 3, is a contradiction The idea is to use lines 2 and 9 to obtain S via *MP*, and thus the contradiction.

Unfortunately, *MP* cannot apply to lines 2 and 9. *MP* requires (1) a conditional sentence and (2) the antecedent of that conditional. The sentence on line 9, $Q \vee R$, is not the antecedent of the conditional on line 2, $(R \vee Q) \supset S$. $Q \vee R$ and $R \vee Q$ are distinct $\mathcal{L}_s$ sentences.

In English, we can reverse the order of the disjuncts in a disjunctive sentence without affecting the sentence's truth conditions. The sentences below mean the same:

> Socrates is wise or Socrates is happy.
> Socrates is happy or Socrates is wise.

The same *commutative* principle holds for disjunctions in $\mathcal{L}_s$. We can reverse the order of the disjuncts in a disjunctive $\mathcal{L}_s$ sentence, thereby producing a new sentence with the same truth conditions as the original

> $W \vee H$
> $H \vee W$

Conjunction in English and in $\mathcal{L}_s$ exhibits the same property. In both English and $\mathcal{L}_s$, reversing the order of a conjunction's conjuncts has no effect on the truth conditions of the sentence as a whole.

Socrates is wise and Socrates is happy.
Socrates is happy and Socrates is wise.

The $\mathcal{L}_s$ counterparts of these sentences are also logically equivalent; they have the same truth conditions

$$W \wedge H$$
$$H \wedge W$$

We can make these remarks systematic by introducing a transformation rule—the *commutative rule (Com)*—permitting the reversal of conjuncts and disjuncts

$$(Com) \quad p \wedge q \dashv\vdash q \wedge p$$
$$p \vee q \dashv\vdash q \vee p$$

Com permits the reversal of conjunct and disjunct pairs, atomic or nonatomic. Applying *Com* in line 10 of the derivation begun earlier, it is now possible to complete the derivation

1.	+	$P \supset Q$	
2.	+	$(R \vee Q) \supset S$	
3.	+	$\neg S$	
4.	?	$\neg(P \vee S)$	
5.		$P \vee S$	
6.		$? \times$	
7.		P	3, 5, $\vee E$
8.		Q	1, 7, MP
9.		$Q \vee R$	8, $\vee I$
10.		$R \vee Q$	9, *Com*
11.		S	2, 10, MP
12.		$S \wedge \neg S$	3, 11, $\wedge I$
13.		P	5–12, IP

Transformation rules are bidirectional. A sentence matching the form on the left-hand side can be transformed into one matching the form on the right, and *vice versa*. Bidirectionality is signalled by the $\dashv\vdash$ symbol, which tells us that the expression on the left deductively yields the expression on the right, and the expression on the right deductively yields the one on the left.

The bidirectionality of transformation rules distinguishes them from derivation rules. Transformation rules are distinguished as well by their application to sentences *within* sentences. A derivation rule like *MP* applies only to whole lines. We could not use *MP* to derive R in the sequence below:

$$\begin{array}{ll} 1. & P \vee (Q \supset R) \\ 2. & Q \\ \vdots & \vdots \end{array}$$

The restriction to whole sentences does not apply to transformation rules. Transformation rules license the replacement of sentences with logically equivalent sentences that have matching truth conditions. The truth functional character of $\mathcal{L}_s$ guarantees that the replacement of a sentential part of a complex sentence with a logically equivalent sentential part has no effect on the truth conditions of the sentence as a whole.

Let us look at some examples. Although the application of *Com* in the sequence below affects only a part of a sentence, it is perfectly acceptable:

$$\begin{array}{lll} 1. & P \supset (Q \vee R) & \\ 2. & P \supset (R \vee Q) & \text{1, } Com \end{array}$$

It would have made no difference if the disjunctive antecedent to which *Com* was applied had been negated

$$\begin{array}{lll} 1. & P \supset \neg(Q \vee R) & \\ 2. & P \supset \neg(R \vee Q) & \text{1, } Com \end{array}$$

The rule is applied to the sentence inside the parentheses. Because $\mathcal{L}_s$ is truth functional, if $Q \vee R$ and $R \vee Q$ have identical truth conditions, then $P \supset \neg(Q \vee R)$ and $P \supset \neg(R \vee Q)$ must have the same truth conditions.

Before turning to derivations in which *Com* figures, let us look at another, related transformation rule, the *associative rule* (*Assoc*).

$$\begin{array}{ll} (Assoc) & p \wedge (q \wedge r) \dashv\vdash (p \wedge q) \wedge r \\ & p \vee (q \vee r) \dashv\vdash (p \vee q) \vee r \end{array}$$

Assoc permits us to slide parentheses back and forth across sequences of $\wedge$s and $\vee$s. Using *Assoc*, we can transform the sentence

$$\neg A \vee (B \vee \neg C)$$

into the sentence

$$(\neg A \lor B) \lor \neg C$$

Like transformation rules generally, *Assoc* can be applied to sentences within sentences. Consider its application in the sequence below:

1. $(M \land (A \land \neg C)) \land (\neg P \supset Q)$
2. $((M \land A) \land \neg C) \land (\neg P \supset Q)$ 1, *Assoc*

We could have applied the rule differently in the case of the second sentence. That is, we could have treated

$$A \land \neg C$$

as an element, the *q* element in the formulation of *Assoc*, and obtained

3. $M \land ((A \land \neg C) \land (\neg P \supset Q))$ 1, *Assoc*

How we apply a rule on an occasion depends on what sentence we want to derive, and that depends on the derivation.

Although transformation rules, unlike rules of inference, are applicable to parts of sentences, they share with inference rules a general restriction: no line can result from the application of more than one rule. Derivations proceed one step at a time. The application of more than one rule, or the application of a single rule more than once, is not permitted in the derivation of a line. This means that a sequence like that below requires *two* steps, *two* applications of *Com*:

1. $+ (P \lor Q) \land R$
2. ? $R \land (Q \lor P)$
3. $R \land (P \lor Q)$ 1, *Com*
4. $R \land (Q \lor P)$ 3, *Com*

Com does not allow the derivation of line 4 from line 1 in a single step. Every line results from a single application of a single rule.

A third transformation rule bears mention in this context. The rule, sometimes called the *principle of tautology* (*Taut*), is formulated below:

$$(Taut) \quad p \dashv\vdash p \land p$$
$$p \dashv\vdash p \lor p$$

Taut expresses the principle that a sentence is logically equivalent to a conjunction or disjunction whose conjuncts or disjuncts consist of that very sentence.

Although the label, *principle of tautology*, is traditional, it is potentially misleading. Appropriately formulated, *every* derivation rule in $\mathcal{L}_s$ expresses a tautology. We might justify the application of the term in this case by noting that the relation depicted in *Taut* is *patently* tautological, whereas those encompassed by other rules are less obvious. If I utter the same sentence twice then, pretty clearly, I have said no more than I would have said had I uttered it just once.

Although the principle of tautology has little application in English, it can figure centrally in derivations in $\mathcal{L}_s$. As with any derivation rule, it applies to *any* sentence of the appropriate form, simple or complex, as the sequence below illustrates:

1. $A \lor A$
2. $\neg(P \supset Q)$
3. $\neg(A \land \neg B) \lor \neg(A \land \neg B)$
4. A 1, *Taut*
5. $\neg(P \supset Q) \land \neg(P \supset Q)$ 2, *Taut*
6. $\neg(A \land \neg B)$ 3, *Taut*

In the construction of derivations, *Taut* can be applied from right to left, as it is in lines 4 and 6 above, or from left to right, as in line 5.

Derivation Heuristics

Single premise derivations:

- In derivations consisting of a single premise, observe the difference between this premise and the desired conclusion, and apply rules so as gradually to diminish this difference.

Start with large differences, and move in the direction of smaller differences, comparing the current line with the conclusion until you achieve a match.

In such cases, "working backwards" from the conclusion to the premise can be especially helpful.

━━━━━━━━━━━━━━Exercises 3.07━━━━━━━━━━━━━━

Construct derivations for the L_s sequences below using
rules discussed thus far, including Com, Assoc, and
Taut. Remember: only one application of a rule per line.

1. $+ P \vee (Q \wedge \neg R)$
   $? (\neg R \wedge Q) \vee P$

2. $+ (P \supset Q) \vee (R \vee S)$
   $? (S \vee R) \vee (P \supset Q)$

3. $+ (P \wedge P) \vee (Q \vee R)$
   $? R \vee (P \vee Q)$

4. $+ (P \vee Q) \wedge (R \wedge S)$
   $? (R \wedge (P \vee Q)) \wedge S$

5. $+ (P \vee (Q \supset R)) \vee (S \vee T)$
   $? ((P \vee S) \vee T) \vee (Q \supset R)$

6. $+ P \supset ((R \vee Q) \supset S)$
   $+ (T \vee S) \supset W$
   $? P \supset (Q \supset W)$

7. $+ \neg (R \wedge T) \supset \neg S$
   $+ (P \vee (Q \wedge Q)) \supset S$
   $? Q \supset R$

8. $+ (P \vee (Q \vee R)) \supset T$
   $+ (S \vee \neg T) \supset R$
   $? T$

9. $+ P \supset Q$
   $+ (R \vee S) \supset \neg Q$
   $? \neg (P \wedge S)$

10. $+ P \supset ((Q \wedge R) \vee S)$
    $+ (R \wedge Q) \supset \neg P$
    $+ T \supset \neg S$
    $? P \supset \neg T$

11. $+ (P \vee Q \supset (R \wedge S)$
    $+ Q$
    $? T \vee R$

12. $+ P \supset R$
    $+ P \vee (R \wedge S)$
    $? Q \vee R$

13. $+ P \supset (S \supset T)$
    $+ S \wedge (Q \vee (\neg T \vee \neg P))$
    $+ Q \supset (S \supset R)$
    $? P \supset (Q \wedge R)$

14. $+ P \vee S$
    $+ S \supset (T \supset P)$
    $+ S \supset T$
    $? R \vee (Q \vee P)$

15. $+ S \supset Q$
    $+ Q \vee (S \wedge T)$
    $? P \vee (Q \wedge Q)$

3.10 TRANSFORMATION RULES: *DEM*

Our next transformation rule takes us to a new level of complexity. The rule, *DeMorgan's Law (DeM)* [after the British logician, Augustus DeMorgan, (1808–1871)], exhibits an important relation between conjunctions and disjunctions. Consider the English conjunction

Socrates is wise and Socrates is happy.

Intuitively, this conjunction is equivalent to the disjunction

It's not the case that either Socrates is
not wise or Socrates is not happy.

Similarly, a disjunction

Socrates is wise or Socrates is happy

is equivalent to an appropriately negated conjunction

It's not the case that both Socrates is
not wise and Socrates is not happy.

Derivation Heuristics

Negation signs:

- Do not be distracted by negation signs in looking
  for appropriate rules: *negation signs will take care
  of themselves.*

Focus on patterns of elementary sentences and
connectives. In the course of a derivation, negation signs
come and go. The odds are good that unless you make a
careless mistake, the derivation of a sentence with the
right connectives and the right atomic constituents in
the right order will result in a sentence with the right
distribution of negation signs.

The negated sentences are stilted and clumsy, but we can see that they mean the same as their more elegant counterparts. This sameness of meaning is captured by DeMorgan's Law:

$$(DeM) \quad p \wedge q \dashv\vdash \neg(\neg p \vee \neg q)$$
$$p \vee q \dashv\vdash \neg(\neg p \wedge \neg q)$$

DeM enables us to move between ∨s and ∧s. Applying *DeM* to $\mathcal{L}_s$ versions of the English sentences above, we obtain the transformational pairs

$$W \wedge H$$
$$\neg(\neg W \vee \neg H)$$

$$W \vee H$$
$$\neg(\neg W \wedge \neg H)$$

Bearing in mind that the occurrence of a negation sign in a rule is to be interpreted as an instruction to *reverse the valence* of the negated expression, *DeM* permits the conversion of a conjunction into a disjunction, or a disjunction into a conjunction, provided (1) the valence of the conjuncts (or disjuncts) is reversed, and (2) the valence of the whole expression is reversed. More pithily still: we can transform a conjunction into a disjunction (or *vice versa*) provided we reverse the valence of (1) each conjunct (or disjunct) and (2) the expression as a whole. Given the disjunction

$$A \vee \neg B$$

DeM licenses its transformation into the conjunction

$$\neg(\neg A \wedge B)$$

The valence of each disjunct is reversed, as is the valence of the whole expression. Applying *DeM* to

$$\neg(P \wedge \neg(Q \vee R))$$

yields the logically equivalent sentence

$$\neg P \vee (Q \vee R)$$

Here, the valence of the conjuncts is reversed—P becomes $\neg P$, and $\neg(Q \vee R)$ becomes $(Q \vee R)$—as is the valence of the entire expression—it is negative in the original sentence and positive in its transformational counterpart. *DeM* could be applied a second time to the portion of the sentence within the parentheses to obtain

$$\neg P \lor \neg(\neg Q \land \neg R)$$

Remember: a derivation of this sentence from the original requires two applications of *DeM*.

─────────────── **Exercises 3.08** ───────────────

Construct derivations for the L_s sequences below using rules discussed thus far, including *DeM*. Remember: only one application of a rule per line.

1. $+ P \lor (Q \land R)$
   $? \neg(\neg P \land (\neg Q \lor \neg R))$

2. $+ \neg(\neg P \lor (\neg Q \land \neg R))$
   $? P \land (Q \lor R)$

3. $+ (\neg P \lor \neg Q) \land (R \land S)$
   $? \neg((\neg R \lor \neg S) \lor (P \land Q))$

4. $+ P \supset \neg(Q \land (R \lor \neg S))$
   $? P \supset (\neg Q \lor (\neg R \land S))$

5. $+ \neg((\neg P \land \neg Q) \lor (\neg R \lor S))$
   $? (P \lor Q) \land (R \land \neg S)$

6. $+ Q \supset S$
   $+ S \supset P$
   $? P \lor \neg Q$

7. $+ P \supset (Q \supset \neg(R \lor S))$
   $+ Q$
   $? P \supset \neg S$

8. $+ P \supset (Q \lor S)$
   $+ \neg S$
   $? \neg P \lor Q$

9. $+ P \supset (Q \lor S)$
   $+ P \supset (R \lor \neg S)$
   $? P \supset (Q \lor R)$

10. $+ P \lor (Q \lor R)$
    $+ Q \supset (R \land S)$
    $? P \lor R$

11. $+ P \land (S \supset \neg Q)$
    $+ P \supset S$
    $? \neg(\neg P \lor Q)$

12. $+ S \lor (Q \supset R)$
    $+ P \lor (Q \land (T \lor \neg R))$
    $? S \lor (T \lor P)$

13. $+ R \supset \neg(\neg P \land Q)$
    $+ R \lor T$
    $+ Q \land (P \supset \neg S)$
    $? S \supset T$

14. $+ S \lor (P \supset (R \supset Q))$
    $+ S \lor P$
    $+ R$
    $? Q \lor S$

15. $+ S \supset (P \land \neg Q)$
    $+ S \lor (P \land R)$
    $? P$

3.11 TRANSFORMATION RULES: *DIST* AND *EXP*

Transformation rules are founded on relations of logical equivalence. They permit the substitution of expressions with the same truth conditions. Because $\mathcal{L}_s$ is a truth functional language, the truth conditions of a sentence remain unaffected when sentential elements are replaced with logically equivalent elements.

Although some transformation rules have obvious parallels in English (*Com* and *DeM* come to mind), others do not. The *distributive rule (Dist)* lacks graceful English applications.

$$(Dist) \quad p \wedge (q \vee r) \dashv\vdash (p \wedge q) \vee (p \wedge r)$$
$$p \vee (q \wedge r) \dashv\vdash (p \vee q) \wedge (p \vee r)$$

Can we find a plausible transformation in English corresponding to *Dist*? Consider the English sentence

Socrates is wise and he is either happy or brave.

The sentence is equivalent to

Socrates is either wise and happy or wise and brave.

The second sentence is awkward, but it possesses the same truth conditions as the original sentence.

Dist licenses moves from conjunctions, in which one conjunct is itself a disjunction, to a disjunction of conjunctions—and *vice versa*; and from disjunctions, in which one disjunct is itself a conjunction, to a conjunction of disjunctions—and *vice versa*. Applying *Dist* to the sentence

$$(\neg A \vee C) \wedge (\neg A \vee \neg D)$$

yields

$$\neg A \vee (C \wedge \neg D)$$

Less obviously, perhaps, *Dist* applied to the sentence

$$(\neg P \supset Q) \wedge (\neg(R \wedge S) \vee \neg T)$$

results in the sentence

$$((\neg P \supset Q) \wedge \neg(R \wedge S)) \vee ((\neg P \supset Q) \wedge \neg T)$$

This example dramatizes the need for practice in the recognition of patterns in $\mathcal{L}_s$. Once you have the hang of it, the application of the rule becomes almost routine. Until that happens, patterns may seem unobvious—or worse, there may seem to be *no* pattern.

Derivation Heuristics

Using *Dist* in derivations

- Consider applying *Dist* when you encounter complex sentences containing repeated elements.

Dist transforms sentences into longer sentences—or into shorter sentences, depending on the direction of application. The sentence

1. $(F \wedge \neg G) \vee (\neg G \wedge (E \supset H))$

with an intermediate application of *Com* yields (or, taken in the opposite direction, is yielded by) the sentence

2. $\neg G \wedge (F \vee (E \supset H))$

Note the presence of shared elements (the $\neg G$) in sentence (1), and the relative lengths of the sentences.

Dist permits the expansion or contraction of sentences containing $\wedge$s and $\vee$s. The *exportation rule* (*Exp*) allows for the manipulation of conditional sentences

$$(Exp) \quad (p \wedge q) \supset r \dashv\vdash p \supset (q \supset r)$$

Exp permits the replacement of a conditional whose antecedent is a conjunction by a conditional with a conditional consequent, and *vice versa*. This principle is at work in our recognition that the English sentences below have the same truth conditions:

> If Socrates is wise and happy then he is brave.
> If Socrates is wise, then, if he is happy, he is brave.

Translated into L_s

$$(W \wedge H) \supset B$$
$$W \supset (H \supset B)$$

More complex applications of *Exp* are illustrated in the sequence below:

1. $(\neg A \wedge B) \supset (C \supset \neg D)$
2. $(P \wedge \neg (Q \vee R)) \supset S$
3. $((\neg A \wedge B) \wedge C) \supset \neg D$ 1, *Exp*
4. $P \supset (\neg (Q \vee R) \supset S)$ 2, *Exp*

The application of *Exp* in line 3 moves from right to left, the application in line 4 moves left to right. If the patterns are not clear to you, try circling components of the complex sentences that correspond to elements in the formulation of the rule.

Exercises 3.09

Construct derivations for the L_s sequences below using rules discussed thus far, including *Dist* and *Exp*. Remember: only one application of a rule per line.

1. $+ \neg P \vee (\neg Q \wedge \neg R)$
   $? \neg ((P \wedge R) \vee (P \wedge Q))$

2. $+ \neg (P \vee Q) \supset R$
   $? \neg P \supset (\neg Q \supset R)$

3. $+ (P \vee Q) \supset (R \supset S)$
   $? ((P \wedge R) \vee (Q \wedge R)) \supset S$

4. $+ P \vee (Q \wedge \neg R)$
   $? (P \vee Q) \wedge \neg (\neg P \wedge R)$

5. $+ \neg (\neg P \vee (\neg Q \wedge \neg R)) \supset S$
   $? P \supset ((Q \vee R) \supset S)$

6. $+ P \supset (Q \wedge R)$
   $+ Q \supset (R \supset S)$
   $? P \supset S$

7. $+ P \supset (S \wedge Q)$
   $+ S \supset R$
   $? (\neg P \vee Q) \wedge (\neg P \vee R)$

8. $+ (P \wedge R) \supset Q$
   $+ P \supset R$
   $? P \supset Q$

9. $+ (P \wedge Q) \supset R$
   $+ (Q \wedge R) \supset S$
   $? (P \wedge Q) \supset S$

10. $+ (S \vee T) \supset (\neg P \vee \neg R)$
    $+ S \vee (Q \wedge T)$
    $? P \supset \neg R$

11. $+ S \supset (P \vee (Q \wedge R))$ 12. $+ S$
    $+ (P \vee R) \supset ((P \vee Q) \supset R)$ $+ \neg R \supset T$
    $? S \supset R$ $? (R \vee S) \wedge (R \vee T)$

13. $+ Q \supset (T \supset (S \supset P))$ 14. $+ P \supset (Q \supset R)$
    $+ (S \supset P) \supset R$ $+ S \supset (P \wedge Q)$
    $+ S \supset (Q \wedge T)$ $+ (S \supset R) \supset T$
    $? R$ $? P \supset T$

15. $+ ((P \wedge Q) \supset R) \supset S$
    $+ T \supset (P \supset (Q \supset R))$
    $? T \supset ((S \vee P) \wedge (S \vee R))$

3.12 RULES FOR CONDITIONALS: *CONTRA* AND *COND*

Conditional sentences can be *contraposed* without affecting their truth
conditions. The contrapositive of a conditional sentence is a sentence in
which the conditional's antecedent and consequent position and valence are
reversed. The English conditional sentence below is paired with its
contrapositive:

If Socrates is wise, then he is happy.
If Socrates is not happy, then he is not wise.

The very same principle is at work in the rule for *contraposition* (*Contra*):

$$(\textit{Contra}) \quad p \supset q \dashv\vdash \neg q \supset \neg p$$

Applying *Contra* to $\mathcal{L}_s$ versions of the sentences above:

$$W \supset H$$
$$\neg H \supset \neg W$$

The sequence below features applications of *Contra* to complex sentences:

1. $\neg A \supset (B \wedge \neg C)$
2. $\neg (P \supset Q) \supset \neg (R \vee \neg S)$

$\quad$ 3. $\neg(B \wedge \neg C) \supset A$ 1, *Contra*
$\quad$ 4. $(R \vee \neg S) \supset (P \supset Q)$ 2, *Contra*

$\quad$ *Conditional equivalence* (*Cond*) provides another rule useful for transforming and manipulating conditional sentences.

$$(Cond) \quad p \supset q \dashv\vdash \neg p \vee q$$

Cond licenses moves between $\supset$s and $\vee$s. The rule allows for the substitution of a $\supset$ for a $\vee$, or *vice versa*, provided the valence of the sentence to the left of the connective is reversed.

Derivation Heuristics

Conditional Proof (*CP*):

- Proofs featuring conditional conclusions are obvious candidates for *CP*.

But:

- Derivations need not have a conditional conclusion for *CP* to be useful. *CP* can be used to derive conditional sentences that are transformable into other sentences.

Suppose, for instance, you are faced with deriving

$$\neg P \vee Q$$

You might (1) suppose P; (2) derive $P \supset Q$; then (3) use Cond to convert this conditional to $\neg P \vee Q$.

$\quad$ Applications of *Cond* are difficult to illustrate in English because of the way we have characterized conditionals in $\mathcal{L}_s$. If the English sentences below do not appear equivalent, that is because you are giving "if...then..." a sense not equivalent to the sense we give to the $\supset$.

$\quad\quad$ If this substance is acid, then the litmus paper turns red.
$\quad\quad$ Either this substance isn't acid or the litmus paper turns red.

In chapter two, we saw that the differences between conditionals in English and in L_s can be overplayed (see § 2.07). In the present case, if you accept the original conditional sentence, you will accept the claim that either the substance is not acid or the litmus paper turns red. True, the disjunctive sentence leaves open the possibility that the substance is not acid and the litmus paper turns red anyway, but that possibility is left open by the original conditional sentence as well. Three applications of *Cond* are illustrated in the sequence below:

1. $A \supset R$
2. $\neg A \supset (B \wedge \neg C)$
3. $R \vee \neg S$
4. $\neg A \vee R$ 1, *Cond*
5. $A \vee (B \wedge \neg C)$ 2, *Cond*
6. $\neg R \supset \neg S$ 3, *Cond*

The sequence provides examples of applications of *Cond* to both simple and complex sentences and examples of its use in each direction: lines 4 and 5 feature left to right deployments; line 6 is obtained via a right to left deployment.

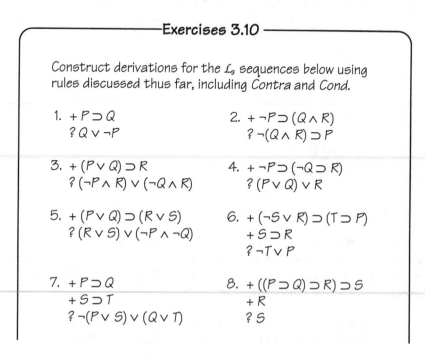

Exercises 3.10

Construct derivations for the L_s sequences below using rules discussed thus far, including *Contra* and *Cond*.

1. $+ P \supset Q$
      $? Q \vee \neg P$

2. $+ \neg P \supset (Q \wedge R)$
      $? \neg (Q \wedge R) \supset P$

3. $+ (P \vee Q) \supset R$
      $? (\neg P \wedge R) \vee (\neg Q \wedge R)$

4. $+ \neg P \supset (\neg Q \supset R)$
      $? (P \vee Q) \vee R$

5. $+ (P \vee Q) \supset (R \vee S)$
      $? (R \vee S) \vee (\neg P \wedge \neg Q)$

6. $+ (\neg S \vee R) \supset (T \supset P)$
      $+ S \supset R$
      $? \neg T \vee P$

7. $+ P \supset Q$
      $+ S \supset T$
      $? \neg (P \vee S) \vee (Q \vee T)$

8. $+ ((P \supset Q) \supset R) \supset S$
      $+ R$
      $? S$

9. $+ (\neg P \supset Q) \supset R$
$+ S \supset (\neg Q \supset P)$
$+ S \vee T$
$? \neg T \supset R$

10. $+ \neg P \vee R$
$+ (P \wedge \neg R) \vee S$
$+ (R \wedge S) \supset Q$
$? P \supset Q$

11. $+ (P \wedge \neg Q) \supset S$
$+ (P \supset Q) \supset \neg T$
$? S \vee \neg T$

12. $+ S \supset R$
$+ R \supset \neg (T \supset Q)$
$? S \supset T$

13. $+ P \vee Q$
$+ Q \supset S$
$+ P \supset S$
$? S \wedge (\neg P \supset Q)$

14. $+ \neg P \supset Q$
$+ S \supset \neg (P \vee Q)$
$+ R \supset S$
$? R \supset T$

15. $+ P \supset (Q \wedge R)$
$+ (P \supset R) \supset S$
$? S$

3.13 BICONDITIONAL SENTENCES: *BICOND*

The rules discussed thus far do not provide a means for coping with derivations featuring biconditionals. Consider the English sequence below:

> The substance is acid if and only if it turns litmus paper red.
> <u>The substance turns litmus paper red.</u>
> Therefore, the substance is acid.

The reasoning appears valid. Suppose we translate the sequence into $\mathcal{L}_s$

1. $+ A \equiv R$
2. $+ R$
3. $? A$

How are we to prove the sequence valid? The sticking point is the biconditional sentence. In chapter two we established that a biconditional sentence is logically equivalent to a conjunction of "back-to-back" conditional sentences. Let us put this logical equivalence to work and introduce a *biconditional equivalence rule* (*Bicond*):

$$(Bicond) \quad p \equiv q \dashv\vdash (p \supset q) \land (q \supset p)$$

Applying the rule to the sequence above, we can now complete the derivation:

$$
\begin{array}{llll}
1. & +\ A \equiv R & & \\
2. & +\ R & & \\
3. & ?\ A & & \\
4. & (A \supset R) \land (R \supset A) & & 1,\ Bicond \\
5. & R \supset A & & 4,\ \land E \\
6. & A & & 2, 5,\ MP \\
\end{array}
$$

By permitting the conversion of biconditional sentences to conjoined conditionals, *Bicond* allows for the elimination of biconditionals in derivations. This reflects our treatment of biconditionals in everyday reasoning. In most contexts, we hear biconditionals (like the biconditional in the English sequence above) as back-to-back conditionals, and reason accordingly.

Exercises 3.11

Construct derivations for the L_s sequences below using rules discussed thus far, including *Bicond*.

1. $+\ P \equiv Q$
   $?\ (P \supset Q) \land (\neg P \supset \neg Q)$

2. $+\ P \equiv Q$
   $?\ (P \lor \neg Q) \land (\neg P \lor Q)$

3. $+\ \neg((P \supset Q) \supset \neg(Q \supset P))$
   $?\ P \equiv Q$

4. $+\ (P \supset Q) \land \neg(\neg P \land Q)$
   $?\ P \equiv Q$

5. $+\ P \equiv Q$
   $?\ Q \equiv P$

6. $+\ (P \lor Q) \supset (R \equiv \neg S)$
   $+\ (S \lor T) \supset (P \land R)$
   $?\ \neg S$

7. $+\ P \lor Q$
   $+\ Q \equiv (R \land S)$
   $+\ (R \lor P) \supset T$
   $?\ T$

8. $+\ P \supset (Q \equiv R)$
   $+\ \neg S \supset (P \lor R)$
   $+\ P \equiv Q$
   $?\ S \lor R$

9. $+\ P \equiv Q$
   $+\ (P \supset R) \supset (P \land S)$
   $?\ \neg(Q \land S) \supset \neg(P \land R)$

10. $+\ S \equiv T$
    $+\ S \supset (P \lor Q)$
    $?\ \neg Q \supset (T \supset P)$

11. $+ P \equiv Q$
 $? (P \wedge Q) \vee (\neg P \wedge \neg Q)$

12. $+ P \supset (Q \equiv R)$
 $+ (\neg Q \vee R) \supset T$
 $? P \supset T$

13. $+ P \equiv (Q \vee R)$
 $+ \neg R \vee \neg S$
 $+ S \supset P$
 $? S \supset Q$

14. $+ P \equiv (\neg Q \vee R)$
 $+ (Q \supset R) \supset S$
 $+ S \supset \neg P$
 $? \neg P$

15. $+ P \vee (Q \wedge R)$
 $+ Q \equiv S$
 $? \neg S \supset P$

3.14 CONSTRUCTIVE DILEMMA (*CD*)

Consider the English sequence below:

> Socrates is wise or he is strong.
> If Socrates is wise, then he is good.
> <u>If Socrates is strong, then he is brave.</u>
> Therefore, Socrates is good or he is brave.

The sequence is valid. We can express it in $\mathcal{L}_s$ as follows (letting $W =$ "Socrates is wise," $S =$ "Socrates is strong," $G =$ "Socrates is good," and $B =$ "Socrates is brave"):

1. $+$ $W \vee S$
2. $+$ $W \supset G$
3. $+$ $S \supset B$
4. $?$ $G \vee B$

We can prove the sequence valid using *IP* and rules already introduced:

1. $+$ $W \vee S$
2. $+$ $W \supset G$
3. $+$ $S \supset B$
4. $?$ $G \vee B$
5. $\quad \lceil \neg(G \vee B)$

6.	$?\times$	
7.	$\neg G \wedge \neg B$	5, DeM
8.	$\neg G$	7, $\wedge E$
9.	$\neg W$	2, 8 MT
10.	$\neg B$	7, $\wedge E$
11.	S	1, 9, $\vee E$
12.	$\neg S$	3, 10, MT
13.	$S \wedge \neg S$	11, 13, $\wedge I$
14.	$G \vee B$	5–13, IP

The pattern of reasoning occurring in the sequence, together with variations on that pattern, is so common that we shall introduce a rule, *constructive dilemma* (*CD*), that permits an inference directly from the premises of the sequence to its conclusion:

$$(CD) \quad p \vee q, p \supset r, q \supset s \vdash r \vee s$$

Applying the rule to the sequence above, we obtain a simple derivation:

1. + $W \vee S$	
2. + $W \supset G$	
3. + $S \supset B$	
4. ? $G \vee B$	
5. $G \vee B$	1, 2, 3, CD

CD is a rule of inference, not a transformation rule. It cannot be applied to sentences within sentences, then, but only to whole sentences at a time. Judicious application of rule *CD* can result in shorter, less complicated derivations. The derivation above is nine steps shorter than the previous derivation in which *CD* was not used.

The advantage afforded by rule *CD* is not merely that it enables some derivations to be shortened. It allows the construction of derivations that are easier to execute because they are closer to the ways we ordinarily reason. Consider the sequence below:

Either Elvis or Fenton fired the shot.
If Elvis fired the shot, Gertrude and Joe are lying.
If Fenton fired the shot, Callie and Gertrude are lying.
Therefore, Gertrude is lying.

The sequence is valid. If the premises are true, the conclusion must be true as well. How might we prove that the sequence is valid in $\mathcal{L}_s$? Because the conclusion is not a disjunction, it might not occur to you to try rule *CD*. Think of it this way. Suppose a sequence includes a disjunction, $p \lor q$. Suppose, further, that the sequence includes *or implies* conditional sentences, the antecedents of which consist of the respective disjuncts, p and q, and the consequents of which are the same sentence, r; that is, suppose the sequence includes $p \supset r$ and $q \supset r$, or these can be derived from the sequence, perhaps via *CP*. In that case, using rule *CD*, the sentence $r \lor r$, and can be derived. This is equivalent, by *Taut*, to r.

This strategy can be applied to the sequence above. (Let E = "Elvis fired," F = "Fenton fired," G = "Gertrude is lying," J = "Joe is lying," and C = "Callie is lying.")

1. + $E \lor F$		
2. + $E \supset (G \land J)$		
3. + $F \supset (C \land G)$		
4. ? G		
5. $\quad\lceil E$		
6. $\quad\mid ? G$		
7. $\quad\mid G \land J$	2, 5, *MP*	
8. $\quad\lfloor G$	7, $\land E$	
9. $\quad E \supset G$	5–8 *CP*	
10. $\quad\lceil F$		
11. $\quad\mid ? G$		
12. $\quad\mid C \land G$	3, 10, *MP*	
13. $\quad\lfloor G$	12, $\land E$	
14. $\quad F \supset G$	10–13 *CP*	
15. $\quad G \lor G$	1, 9, 14, *CD*	
16. $\quad G$	15, *Taut*	

The use of *CP* together with *CD* can save time and agony in derivations of sequences that might otherwise resist solution.

We have now encountered all the rules we require to construct derivations in $\mathcal{L}_s$. The derivation rules include both rules of inference and transformation rules. Transformation rules govern the replacement of sentences with logically equivalent counterparts. Inference rules differ from transformation rules in applying only to whole lines of derivations. We have supposed that rules are formulated in the metalanguage. Negation signs appearing in rule formulations are taken to refer, not to negation signs in $\mathcal{L}_s$ sentences, but to *valences* of sentences. The transformations and inferences permitted by the derivation rules sometimes require the reversal of valences. Conditional equivalence (*Cond*), for instance, permits the replacement of a conditional sentence, $p \supset q$, with a disjunction, $\neg p \vee q$, provided the valence of the conditional's antecedent is reversed. Thus, nonnegated antecedents take on negation signs, and negated antecedents lose them. This is old hat. I mention it now in order to make clear the point of excluding from our list of rules an explicit rule for double negation.

Derivation Heuristics

Indirect Proof (*IP*):

- Sometimes *IP* affords simpler derivations of conditional sentences than *CP*.

Imagine that you want to derive

$$P \supset Q$$

but you get nowhere using *CP*. You might try (1) supposing $\neg(P \supset Q)$ and; (2) deriving a contradiction.

Note that $\neg(P \supset Q)$ can be transformed into $\neg(\neg P \vee Q)$ via *Cond*. Negated disjunctions are transformable using *DeM* into useful conjunctions; here: $P \wedge \neg Q$. Now each conjunct can be "brought down" by $\wedge E$ and used separately in the derivation.

Breaking sentences down in this way is sometimes helpful when you are stuck in the midst of a derivation.

- When in doubt, break complex sentences into simpler sentences.

Look for conjunctions that can be separated into elements, and for applications of *MP*, *MT*, and $\vee E$ rules that facilitate the breaking down of complex sentences.

Transformation rules may also be useful in establishing subgoals. When faced with a derivation in which the route to the conclusion is not obvious, it may be helpful to think of ways in which the transformation rules could be used to transform the conclusion. It is entirely possible that one of these transformed sentences will be easier to derive. If that is so, then you need only derive the transformed sentence, and then reverse the process to arrive at the conclusion. The strategy is simply a version of the more general strategy of working backward. You can look over the conclusion and work backwards in your head to the premises. In a difficult derivation, this may not take you far. But the aim, as always, is to apply rules to *narrow the gap* between where you are and where you want to be.

As you familiarize yourself with the rules and their application in derivations, patterns will begin to emerge. After practice, the solution of a derivation that once seemed formidable will often be obvious at a glance— even though the derivation that results from a flash of insight may turn out to have many steps. At first there may be no insight, even for the simplest of derivations. In that case, it is often useful to apply rules—for instance, *MP*, *MT*, $\wedge E$, and $\vee E$—that result in simplifying or breaking down complex sentences. Simpler sentences are often easier to manipulate.

The importance of practice has been emphasized, but it cannot be *over*emphasized. The skills required for derivation construction are mostly perceptual. They can be developed only through practice. It is often useful to take a difficult derivation already constructed and *re*construct it on a fresh sheet of paper. This is like playing scales on the piano: It *feels* mindlessly repetitive, but it is in fact part of what it takes to acquire the requisite skill. Eventually, the light dawns and arrays of symbols that once appeared shapeless form themselves into units marching inevitably toward a conclusion.

Exercises 3.12

Prove the validity of each sequence below, incorporating an application of CD.

1. $+ \neg(P \wedge Q)$
   $+ S \supset P$
   $+ S \supset Q$
   $? \neg S$

2. $+ S \supset P$
   $+ Q \supset P$
   $+ \neg Q \supset S$
   $? P$

3. $+ P \supset (R \vee S)$
   $+ Q \supset (R \vee S)$
   $? (P \vee Q) \supset (R \vee S)$

4. $+ P \supset (R \wedge T)$
   $+ Q \supset (S \wedge T)$
   $? (P \vee Q) \supset (R \vee S)$

5. $+ \neg P \supset S$
   $+ \neg R \supset (S \vee T)$
   $+ \neg(P \wedge R)$
   $? S \vee T$

6. $+ P \vee R$
   $+ P \supset (Q \wedge \neg S)$
   $+ (\neg R \vee T) \wedge \neg S$
   $? Q \vee T$

7. $+ \neg P \vee Q$
   $+ R \supset S$
   $? (P \vee R) \supset (Q \vee S)$

8. $+ \neg P$
   $+ \neg Q$
   $? (P \vee Q) \supset (R \vee S)$

9. $+ P \supset (S \vee T)$
   $+ Q \supset (S \vee T)$
   $+ \neg T$
   $? (P \vee Q) \supset S$

10. $+ P \supset (Q \vee R)$
    $+ S \supset (R \vee T)$
    $+ \neg R$
    $? (P \vee S) \supset (Q \vee T)$

11. $+ S \vee \neg T$
    $+ P \vee \neg Q$
    $? (Q \vee T) \supset (S \vee P)$

12. $+ P \vee Q$
    $+ R \vee S$
    $? \neg(Q \wedge S) \supset (P \vee R)$

13. $+ P \vee Q$
    $+ P \supset (R \vee S)$
    $+ \neg Q \vee (S \vee R)$
    $? R \vee S$

14. $+ S \supset T$
    $+ R \supset (T \vee Q)$
    $+ (T \vee Q) \supset P$
    $? (S \vee R) \supset (T \vee P)$

15. $+ P \supset S$
    $+ Q \supset S$
    $? (P \wedge R) \vee (Q \wedge R)) \supset S$

3.15 PROVING INVALIDITY

The reasoning we encounter in everyday life is often flawed in one way or another. Sometimes unwarranted assumptions contaminate an argument so that, even when its conclusion is validly supported, we have little reason to accept that conclusion. Sometimes arguments are out and out invalid. The premises of an argument may be plausible, and the conclusion a sentence we would like to believe, but the conclusion may fail to follow from the premises.

Derivations provide proofs for the validity of sequences. What of invalid sequences? If you fail to prove a sequence valid, you do not thereby prove it invalid. There may *be* a proof for validity that you have overlooked. Imagine that you cannot find such a proof. You suspect that the sequence is invalid, but how can you prove that it is?

One way to establish that a sequence is invalid is by means of a truth table (see § 3.02). Bearing this in mind, you might construct a truth table and see whether, on any row of that truth table, the premises of the suspicious sequence are true and its conclusion is false. The strategy lacks appeal. Truth tables, though reliable, are unwieldy and potentially confusing when they require many rows.

Happily, an alternative is available. A sequence is invalid if an assignment of truth values to its atomic components results in its premises being true and its conclusion false. Suppose that given a sequence $\langle P,c \rangle$, we assigned values to c to make c false, and then sought a consistent assignment of values to P to make P true. Were we to find such an assignment, the sequence would be proved invalid. Best of all, the proof would have been accomplished without the construction of a truth table.

Consider the sequence below:

1. $+ P \supset Q$
2. $+ R \supset S$
3. $+ R \vee Q$
4. ? $P \vee S$

Suppose you have tried without success to construct a derivation to prove this sequence valid, and you suspect that it is invalid. You could construct a truth table to confirm your suspicions

P Q R S	$P \supset Q$	$R \supset S$	$R \lor Q$	$P \lor S$
T T T T	T	T	T	T
T T T F	T	F	T	T
T T F T	T	T	T	T
T T F F	T	T	T	T
T F T T	F	T	T	T
T F T F	F	F	T	T
T F F T	F	T	F	T
T F F F	F	T	F	T
F T T T	T	T	T	T
F T T F	T	F	T	F
F T F T	T	T	T	T
F T F F	T	T	T	F
F F T T	T	T	T	T
F F T F	T	F	T	F
F F F T	T	T	F	T
F F F F	T	T	F	F

The truth table reveals an assignment of truth values to the sequence's atomic sentences that shows it to be invalid; that assignment occurs in the fifth row up. In that row, the premises of the sequence are true and its conclusion is false.

Let us look at a streamlined technique for achieving the same end. The technique enables us to discover single rows of truth tables that establish the invalidity of invalid sequences. A sequence is invalid if there is an interpretation—an assignment of truth values to its atomic constituents—that makes its premises true and its conclusion false. We first find an assignment that makes the conclusion false, and then determine whether this assignment is can be consistently extended to the premises to make them true.

The procedure is illustrated below. First, set out the sequence horizontally, with premises to the left and the conclusion to the right, separated from the premises by a vertical line

$$P \supset Q \quad R \supset S \quad R \lor Q \mid P \lor S$$

Second, assign values to the atomic sentences making up the conclusion to make the conclusion false. In this instance the conclusion consists of a disjunction, so there is only one way to do this—by assigning the value *false* to both constituents, P and S

$$P \supset Q \quad R \supset S \quad R \vee Q \mid P \vee S$$
$$ \text{F} \ \ \text{F}$$
$$ \underset{\smile}{}$$
$$ \text{F}$$

Assignments of truth values must be consistent, so these values must be carried over to the premises

$$P \supset Q \quad R \supset S \quad R \vee Q \mid P \vee S$$
$$\text{F} \text{F} \text{F} \ \ \text{F}$$
$$ \underset{\smile}{}$$
$$ \text{F}$$

Third, seek truth value assignments to the remaining premises that result in their being true. In the present case this means that R must be false—otherwise the second premise would be false. If R is false, Q must be true if the third premise is to be true. These assignments result in the premises' being true

$$P \supset Q \quad R \supset S \quad R \vee Q \mid P \vee S$$
$$\text{F} \ \ \text{T} \ \ \text{F} \quad \text{F} \ \ \text{F} \quad \text{F} \ \ \text{T} \quad \ \ \text{F} \ \ \text{F}$$
$$\underset{\smile}{} \quad \ \ \underset{\smile}{} \quad \ \ \underset{\smile}{} \quad \ \ \underset{\smile}{}$$
$$\ \ \text{T} \quad \ \ \ \text{T} \quad \ \ \ \text{T} \quad \ \ \ \ \ \text{F}$$

We have uncovered a consistent assignment of truth values, an interpretation, under which the premises of the sequence are true and its conclusion is false; I: $\{P=F, Q=T, R=F, S=F\}$. Compare this interpretation to the fifth row up in the original truth table. There, P, Q, R, and S have just these values. The technique outlined above manufactures *single rows* of truth tables; these rows establish the invalidity of invalid sequences.

This technique can be extended to any $\mathcal{L}_s$ sequence. When it proves successful and we can discover a consistent assignment of truth values that makes the conclusion of a sequence false and its premises true, we have a tidy proof of invalidity. Failure to discover such an assignment, of course, does not establish the validity of a sequence. Just as we may fail to discover a proof for a perfectly valid sequence, so we may fail to find an interpretation for an invalid sequence that proves its invalidity, which would be revealed were we to construct a truth table. There is often more than one way to assign values to a conclusion to make it false or to a premise to make it true.

It may be necessary to test several interpretations before abandoning an attempt to prove invalidity and embarking on a proof for validity.

Consider the sequence below:

1. $+ P \wedge (Q \supset R)$
2. $+ (R \supset Q) \supset P$
4. ? $P \wedge (Q \equiv R)$

which we could represent horizontally

$$P \wedge (Q \supset R) \quad (R \supset Q) \supset P \mid P \wedge (Q \equiv R)$$

Several assignments of truth values to the conclusion of this sequence render it false. Some of these assignments will, when carried over into the premises, result in false premises as well. Thus, we can make the conclusion false by assigning the value *false* to *P*, and *true* to both *Q* and *R*

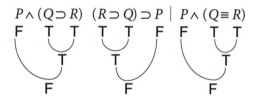

That assignment results in both premises being false. Were we to stop here and assume that because *this* assignment fails to establish invalidity, *no* assignment will do so, we would be in error. As the diagram below illustrates, other assignments must be considered:

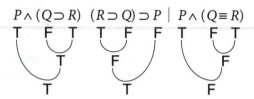

Here is an assignment of truth values, an interpretation, under which the conclusion of the sequence is false and its premises are true

$$I: \{P=T, Q=F, R=T\}$$

Exercises 3.13

For each of the sequences below (1) construct a derivation proving its validity, or (2) provide an interpretation demonstrating its invalidity and a chart showing the application of this interpretation.

1. $+ \neg(P \wedge Q) \supset ((R \supset S) \supset T)$
   $+ R \wedge (P \vee Q)$
   $? S \supset T$

2. $+ P \equiv (Q \vee R)$
   $+ \neg Q$
   $? \neg P \supset R$

3. $+ P \equiv \neg Q$
   $+ (Q \wedge R) \supset (S \vee T)$
   $+ \neg(P \wedge R)$
   $? (R \supset \neg S) \supset T$

4. $+ (P \vee Q) \supset (R \equiv S)$
   $+ \neg(\neg S \wedge P)$
   $+ R \supset T$
   $? P \supset (T \wedge R)$

5. $+ P \supset (Q \supset R)$
   $+ R \supset S$
   $+ (T \wedge U) \supset P$
   $+ \neg(\neg U \wedge \neg Q)$
   $? T \supset (Q \wedge S)$

6. $+ \neg P \supset (Q \supset R)$
   $+ (P \vee S) \supset T$
   $+ R \supset (P \vee S)$
   $+ \neg T$
   $? \neg Q$

7. $+ \neg(P \wedge Q) \supset R$
   $+ R \supset S$
   $+ (P \wedge S) \supset Q$
   $? \neg R \vee Q$

8. $+ P \supset (Q \equiv R)$
   $+ \neg S \supset (P \vee R)$
   $+ P \equiv Q$
   $? S \wedge R$

9. $+ \neg(P \wedge Q)$
   $+ (P \supset R) \supset (P \wedge S)$
   $? (P \wedge R) \supset (Q \wedge S)$

10. $+ P \supset (Q \supset R)$
    $+ \neg R$
    $? P \supset \neg Q$

11. $+ P \supset (Q \vee R)$
    $+ S \supset (T \vee R)$
    $? (P \vee S) \supset ((T \vee Q) \vee R)$

12. $+ P \supset (Q \vee R)$
    $+ S \supset (T \vee R)$
    $? (P \vee S) \supset R$

13. $+ R \supset (P \vee Q)$
    $+ S \supset (P \vee Q)$
    $+ P \vee Q$
    $? R \wedge S$

14. $+ (P \vee Q) \supset R$
    $+ (P \vee Q) \supset S$
    $+ \neg S$
    $? \neg P$

15. $+ R \supset Q$
    $+ P \supset (R \vee S)$
    $? P \wedge Q$

3.16 THEOREMS IN $\mathcal{L}_s$

Formal systems yield *theorems*. Theorems are sentences that follow exclusively from axioms of the system without any further assumptions. When we prove theorems in Euclidean geometry, we derive geometrical sentences from Euclidean axioms. Once a theorem is shown to follow from an axiom, we are free to use it as a premise in other derivations because it is always possible to trace such derivations back to the axioms on which the system is based. A theorem like the Pythagorean theorem

> The square of the hypotenuse of a right triangle =
> the sums of the squares of the two remaining sides

requires no assumptions other than those concerning points, lines, and angles contained in the Euclidean axioms.

Suppose, in contrast, you set out to determine the length of the hypotenuse of a right triangle formed by the intersection of three highways in Nebraska. In this case, you require, in addition to the axioms of geometry, facts concerning the angle of each intersection and the distances between intersections. The answer you obtain is not a theorem of geometry, but a purportedly fact-stating sentence about the world. Geometers prove theorems, and carpenters and land surveyors use geometry to establish actual areas and boundaries.

The axioms of a system determine the character of the system. Euclidean geometry has five axioms from which every Euclidean theorem is derivable. The axioms are geometrical sentences that occupy a privileged position within the system: they are taken not to require proof. For centuries geometers imagined that the Euclidean axioms were definitive of the structure of space and that they had a *natural* basis. The geometers were wrong. Space, as we now know, is not Euclidean. Moreover, we no longer suppose that the axioms of a formal system need worldly support. If they are properly formulated, they define a self-contained system that may or may not have an application to the world.

Like geometry, arithmetic constitutes a formal system. Familiar arithmetical truths like

$$7 + 5 = 12$$

express arithmetical theorems. These can be derived from five axioms of arithmetic that have received various formulations in this century. Although both arithmetic and Euclidean geometry are founded on five axioms, there is nothing magical about this number. A formal system might feature any number of axioms. $\mathcal{L}_s$ represents a limiting case: it possesses *no* axioms. It is possible to devise an equivalent, axiom-based system. We opted rather to create $\mathcal{L}_s$ as a *natural deduction* system. The system includes no axioms and so, in one sense, it resembles natural reasoning. When we produce arguments in ordinary life, we appeal to premises, but never—or rarely—to sentences that have the character of axioms. So it is with $\mathcal{L}_s$.

Derivation Heuristics

- Derivations of theorems involving conditionals are sometimes simpler using IP rather than CP.

Consider the theorem below:

$$\vdash P \supset (Q \supset P)$$

It might occur to you to derive the theorem using an embedded CP:

1. $\quad P$
2. $\quad ?\ Q \supset P$
3. $\quad\quad Q$
4. $\quad\quad ?\ P$

The theorem can be proved in this way—for instance, by obtaining p via IP—but a simpler IP derivation is possible:

1.	$\neg(P \supset (Q \supset P))$	
2.	$?\ \times$	
3.	$\neg(\neg P \vee (Q \supset P))$	1, Cond
4.	$\neg(\neg P \vee (\neg Q \vee P))$	3, Cond
5.	$P \wedge \neg(\neg Q \vee P)$	4, DeM
6.	$\neg(\neg Q \vee P)$	5, $\wedge$E
7.	$Q \wedge \neg P$	6, DeM
8.	$\neg P$	7, $\wedge$E
9.	P	5, $\wedge$E
10.	$P \wedge \neg P$	8, 9, $\wedge$I
11.	$P \supset (Q \supset P)$	1–10, IP

Despite lacking axioms, $\mathcal{L}_s$ has theorems. Theorems are sentences belonging to a formal system that follow from the axioms of the system alone. How then could $\mathcal{L}_s$ be thought to yield theorems? If a theorem follows from the axioms of a system and $\mathcal{L}_s$ has no axioms, then a theorem of $\mathcal{L}_s$ is just a sentence that follows from *any set of sentences whatever*, including the *empty set* of sentences. Does this make sense? Consider the sentence

$$P \vee \neg P$$

Now, look at a derivation of this sentence

1. $\quad \lceil \neg(P \vee \neg P)$
2. $\quad \mid ? \times$
3. $\quad \lfloor \neg P \wedge P$ 1, *DeM*
4. $\quad P \vee \neg P$ 1–3, *IP*

The derivation leads off with a suppositional premise: suppose that $P \vee \neg P$ is false. This leads to a contradiction, so $P \vee \neg P$ must be true. To indicate that a sentence is to be taken as a theorem, we can deploy the turnstyle, $\vdash$. The sentence above would be written as follows:

$$\vdash P \vee \neg P$$

The $\vdash$ is used to express the derivability relation. Given a sequence, $\langle P, c \rangle$, if $P \vdash c$, then c is derivable from P, that is, a derivation of c from P is possible. If a sentence, p, is a theorem, then $\vdash p$; p is derivable from the empty set of sentences. The first step in the derivation of a theorem must be a suppositional step: theorems are proved using either *IP* or *CP*, or some *IP/CP* hybrid.

Let us consider a more complex theorem

$$\vdash (P \supset Q) \equiv (\neg P \vee Q)$$

The theorem is a biconditional. We derive it by treating it as a two-way conditional. We first derive $(P \supset Q) \supset (\neg P \vee Q)$; next we derive $(\neg P \vee Q) \supset (P \supset Q)$; finally we put these together to form the biconditional

1. $\quad \lceil P \supset Q$
2. $\quad \mid ? \neg P \vee Q$
3. $\quad \lfloor \neg P \vee Q$ 1, *DeM*

4.	$(P \supset Q) \supset (\neg P \vee Q)$	1–3, CP
5.	$\lceil \neg P \vee Q$	
6.	$\mid ? P \supset Q$	
7.	$\lfloor P \supset Q$	5, DeM
8.	$(\neg P \vee Q) \supset (P \supset Q)$	5–7, CP
9.	$((P \supset Q) \supset (\neg P \vee Q)) \wedge ((\neg P \vee Q) \supset (P \supset Q))$	4, 8, $\wedge$I
10.	$(P \supset Q) \equiv (\neg P \vee Q)$	9, Bicond

This completes our introduction to the elements of $\mathcal{L}_s$. The next section provides a discussion of proofs *about* $\mathcal{L}_s$. Before venturing into that discussion, you may find it useful to review §§ 2.19–2.21.

Exercises 3.14

Construct derivations for the sentences below proving that they are theorems of $\mathcal{L}_s$.

1. $\vdash \neg(P \wedge \neg P)$

2. $\vdash P \supset (\neg P \supset P)$

3. $\vdash (P \wedge \neg P) \supset Q$

4. $\vdash ((P \supset Q) \wedge \neg Q) \supset \neg P$

5. $\vdash ((P \supset Q) \wedge P) \supset Q$

6. $\vdash (P \vee Q) \equiv \neg(\neg P \wedge \neg Q)$

7. $\vdash P \supset P$

8. $\vdash (P \supset Q) \vee (Q \supset P)$

9. $\vdash P \supset (Q \supset P)$

10. $\vdash \neg(P \supset Q) \equiv (P \wedge \neg Q)$

11. $\vdash (P \supset Q) \equiv (\neg P \vee Q)$

12. $\vdash \neg P \supset (P \supset Q)$

13. $\vdash P \supset ((P \wedge Q) \vee (P \wedge \neg Q))$

14. $\vdash (P \supset Q) \supset (P \supset (Q \equiv P))$

15. $\vdash ((P \supset Q) \supset R) \supset ((P \supset Q) \supset (P \supset R))$

3.17 SOUNDNESS AND COMPLETENESS OF $\mathcal{L}_s$

Theorems of $\mathcal{L}_s$ are sentences derivable from the empty set of sentences. All of the theorems we have encountered thus far are logical truths: they are true under any interpretation—true in all possible worlds. But wait: can we really be sure that every provable theorem expresses a logical truth? More generally, what assurance do we have that every derivable sequence producable in $\mathcal{L}_s$ is a valid sequence? These questions concern the *soundness* of $\mathcal{L}_s$. We can define *soundness* as follows:

> $\mathcal{L}_s$ is *sound* if and only if, for every set of sentences, P, and any sentence, c, if P *deductively yields* c ($P \vdash c$), then P *logically implies* c ($P \vDash c$).

If P deductively yields c ($P \vdash c$), then there is a derivation from P to c that uses the derivation rules we have established for $\mathcal{L}_s$. If P logically implies c ($P \vDash c$), then there is no interpretation under which P is true and c false. If $\mathcal{L}_s$ is sound, then every derivable sequence is a valid sequence.

What of the complementary question: is every valid sequence derivable? This question concerns the *completeness* of $\mathcal{L}_s$.

> $\mathcal{L}_s$ is *complete* if and only if, for every set of sentences, P, and any sentence, c, if $P \vDash c$, then $P \vdash c$.

Thus, $\mathcal{L}_s$ is complete just in case, corresponding to every valid sequence expressible in $\mathcal{L}_s$, there is a derivation establishing its validity and relying only on the rules set out in this chapter. This is not to say that we can always *find* such a derivation, only that if we are unable to devise a derivation, the fault is ours, not $\mathcal{L}_s$'s.

We can arrive at an intuitive understanding of the notions of soundness and completeness by considering the relation between the concepts of validity and logical implication on the one hand, and the concept of derivability on the other hand. A derivation in $\mathcal{L}_s$ is a finite ordered sequence of sentences, $\langle P,c \rangle$, in which one sentence, c (the conclusion), is derivable from the other sentences, P (the premises), by means of one or more derivation rules. We have symbolized c's derivability from P (alternatively, P's yielding c) via a turnstile: $P \vdash c$. Theorems are degenerate cases of derivations. A theorem, t, is a sentence derivable from the empty set of sentences: $\vdash t$.

How are derivability and validity related? The concept of derivability is a *syntactic* notion. Derivation rules permit us to add sentences to a sequence given syntactic features of other sentences present in the sequence regardless of their interpretation. Validity, in contrast, is a *semantic* concept. We define validity by reference to *truth* and the derivative notion of an interpretation. A sequence is valid when its premises logically imply its conclusion; and the premises of a sequence logically imply its conclusion when there is no interpretation under which the premises are true and the conclusion is false. In establishing the soundness and completeness of $\mathcal{L}_s$, we establish a correspondence between syntactic and semantic categories.

If you have been observant, you may have noticed that some of the rules we have used provide convenient short cuts, but they could otherwise be omitted. Consider $\vee E$

$$(\vee E) \quad p \vee q, \neg p \vdash q$$

We could omit this rule and still get by in derivations. Imagine that we had a sequence of the sort below:

1. $+ A \vee B$
2. $+ \neg A$
3. $?\ B$

We need only convert the disjunction to a conditional, and apply *MP*

1. $+ A \vee B$
2. $+ \neg A$
3. $?\ B$
4. $\quad \neg A \supset B \qquad\qquad$ 1, Cond
5. $\quad B \qquad\qquad\qquad$ 2, 4, *MP*

We can distinguish indispensible rules from those that we could, at the cost of some inconvenience, do without. An indispensible rule is a *primitive rule*; the remaining rules are *derived rules*. Which of the rules set out for $\mathcal{L}_s$ are primitive, and which are derived? In part, this is a matter of choice. The sequence above shows how we can dispense with $\vee E$, using instead *Cond* and *MP*. We might just as well have dispensed with *MP*: where we now apply *MP*, we could instead apply *Cond* and $\vee E$. We can settle on a set of primitive rules by selecting a set of rules none of which is replaceable by other rules (or combinations of rules) in the set. Derived rules can be defined relative to this set of primitive rules as those replaceable in derivations by one or more primitive rules.

In assessing the soundness of $\mathcal{L}_s$, we must determine whether the primitive rules are truth preserving, whether the application of primitive rules to true sentences always yields true sentences. Because derived rules constitute shortcuts, which are themselves derivable via applications of primitive rules, it follows that if the primitive rules are truth preserving, L_s is sound. L_s is complete, provided that every valid sequence is provably valid by means of a derivation that uses only primitive derivation rules.

Primitive and Derived Rules

A derived rule is replaceable by applications of one or more primitive rules. We have seen that there is some latitude in choosing which rules are to be primitive, and which derived. How much latitude?

A set of rules for L_s must be adequate to yield derivations of every valid sequence expressible in L_s. It is also desirable for the set of primitive rules to be the smallest possible, that is, to contain no rules that are derivable from other rules.

In L_s, we can take these rules as primitive, the rest as derived:

$\wedge$I and $\wedge$E; $\vee$I and $\vee$E; MP; IP; CP; and Bicond.

The remaining rules, including all the transformation rules, can be derived from these—assuming that negation signs in the formulation of rules are to be read as indicators of the relative valences of sentences. The assumption enables us to dispense with a rule for double negation—that is, a rule that licenses an inference from $\neg\neg p$ to p.

It is worth reflecting on what proofs for the soundness and completeness of L_s would require, even though they will not be attempted here. First, note that such proofs are constructed in the metalanguage: they are not proofs *in* L_s, they are proofs *about* L_s. This is no accident. Proofs for soundness or completeness of any formal system must be formulated in the metalanguage.

What would a proof for soundness involve? A sequence is valid if its premises logically imply its conclusion. If a set of sentences, *P*, logically implies a sentence, c, then, if P is true, c must be true. Suppose the premises of an arbitrary sequence are true. If we could show that *if a sentence is true, then any sentence derived from those sentences by means of a primitive derivation rule is true*, then we would have shown that L_s is sound. This would provide an *inductive proof* for the soundness of L_s (not to be confused with ordinary inductive reasoning; see the box at the end of the chapter). If a rule permits the derivation of only true sentences from true sentences, it is *truth preserving*. Showing that L_s is sound is a matter of showing that each of the derivation rules is truth preserving in this sense.

With two exceptions, *CP* and *IP*, we can use truth tables to establish that the primitive derivation rules of L_s are truth preserving. Because of their form, *CP* and *IP* require special treatment. Although this complicates the proof, it does not change it in any fundamental way.

A proof that L_s is complete is more complicated, even in outline. We must first show that every logical truth expressible in L_s is derivable as a theorem. We then show that every valid sequence is derivable.

Consider the truth table for an arbitrary sentence expressible in L_s, $\neg P \wedge Q$

P Q	$\neg P$	$\neg P \wedge Q$
T T	F	F
T F	F	F
F T	T	T
F F	T	F

This sentence *or its negation*, $\neg(\neg P \wedge Q)$, is derivable from premises consisting of its atomic constituents, negated or not, depending on whether those constituents are false or true on a given row of the truth table. Going row by row

$$P, Q \vdash \neg(\neg P \wedge Q)$$
$$P, \neg Q \vdash \neg(\neg P \wedge Q)$$
$$\neg P, Q \vdash (\neg P \wedge Q)$$
$$\neg P, \neg Q \vdash \neg(\neg P \wedge Q)$$

What is true for the sentence in this example turns out to be provably true for *every* sentence of L_s. That is, every sentence expressible in L_s or its negation is derivable from premises consisting of its atomic constituents, negated or not, depending on whether those constituents are false or true on a given row of the sentence's truth table.

Now consider the logical truths. A logical truth is a sentence true under every interpretation. A logically true sentence, $P \supset (Q \supset P)$, for instance, has the value "true" in every row of its truth table. It can be shown that if this is so, then $P \supset (Q \supset P)$ is derivable whatever the values of P and Q

$$P, Q \vdash P \supset (Q \supset P)$$
$$P, \neg Q \vdash P \supset (Q \supset P)$$
$$\neg P, Q \vdash P \supset (Q \supset P)$$
$$\neg P, \neg Q \vdash P \supset (Q \supset P)$$

If $P \supset (Q \supset P)$ is so derivable, $P \supset (Q \supset P)$ is derivable as a theorem of $\mathcal{L}_s$. More generally, every logical truth is derivable as a theorem: if p is implied by any sequence, including the empty sequence ($\vDash p$), then p is derivable from any sequence including the empty sequence ($\vdash p$).

Although the details are complicated, given this result, it can be shown that if $P \vDash c$, then $P \vdash c$. The intuitive idea is as follows. Every sequence has a "corresponding conditional." Consider the sequence

1. $+ \ \neg P \supset Q$
2. $+ \ \neg Q$
3. ? P

The sequence is valid, so

$$\neg P \supset Q, \neg Q \vDash P$$

We can construct a conditional sentence from this sequence, replacing the $\vDash$ with a $\supset$, and conjoining the premises

$$((\neg P \supset Q) \wedge \neg Q) \supset P$$

If the original sequence is valid, then its corresponding conditional must be logically true. (Why? Well, the conditional is false only if its antecedent— here the conjoined premises of the original sequence—is true and its consequent—the conclusion of the original sequence—is false. If the sequence is valid, this is impossible.) If every valid sequence corresponds to a logically true conditional, and if every logically true sentence is derivable, then the conditional corresponding to every valid sequence is derivable. And if that is so, the sequence is derivable.

These brief remarks on soundness and completeness barely scratch the surface. The aim is to suggest what is involved in a proof that $\mathcal{L}_s$ has these properties. More detailed discussions of the proofs can be found in Geoffrey Hunter's *Metalogic: An Introduction to the Metatheory of Standard First Order Logic* (Berkeley: University of California Press, 1971); S. C. Kleene's *Mathematical Logic* (New York: John Wiley and Sons, 1967); and Paul Teller's *A Modern Formal Logic Primer: Predicate Logic and Metatheory* (Englewood Cliffs: Prentice-Hall, 1989).

Mathematical Induction

Suppose you want to show that every natural number has a particular property, ϕ. (A natural number is one of the numbers 0, 1, 2, 3, 4, etc.)

You could do so if you could show (1) that 0 has ϕ, and (2) that if a number has ϕ, then its successor, the number following it in the series, has ϕ. Thus: if 0 has ϕ, then the successor of 0, 1, has ϕ; if 1 has ϕ then 2, its successor has ϕ; and so on.

This technique, proof by *mathematical induction*, enables us to establish that every member of an unlimited or infinite series of objects has a particular property if the first member of the series has the property and, if a member of the series has it, the next member has it.

Mathematical induction should not be mistaken for what is commonly called *inductive reasoning*. Ordinary inductive reasoning is distinguished from *deductive* reasoning. When we reason deductively, the truth of our premises guarantees the truth of our conclusion. From "All people are mortal" and "Socrates is a person," we can infer "Socrates is mortal." When we reason inductively, our premises provide only probabilistic support for our conclusion. From "80% of country club members are Republicans" and "Chip is a member of a country club," we can infer that "It is likely that Chip is a Republican." In so doing, we are reasoning inductively.

Mathematical induction, then, is a species of deductive reasoning!

4

The Language $\mathcal{L}_p$

Asentential logic like $\mathcal{L}_s$ is a blunt instrument. Although $\mathcal{L}_s$ captures an important range of logical relations among sentences, its limitations quickly become evident when we consider sequences like the following:

> All people are mortal.
> <u>Socrates is a person.</u>
> Therefore, Socrates is mortal.

The sequence is clearly valid; its conclusion follows from its premises. When we represent it in $\mathcal{L}_s$, however (letting P = "All people are mortal," S = "Socrates is a person," and M = "Socrates is mortal"), its validity is masked

> 1. + P
> 2. + S
> 3. ? M

Finding an interpretation under which the conclusion, M, is false and the premises, P and S, are true is simple: assign the value *true* to P and S, and *false* to M. The example illustrates one respect in which $\mathcal{L}_s$ is logically impoverished: it cannot reflect logical features *internal* to atomic sentences.

The class of valid proofs in $\mathcal{L}_s$ includes only those whose validity is determined by the truth functional structure of sentences. As we have just seen, not all logical relations are like this. This had been known since the time of Aristotle, but not until the late nineteenth century did Gottlob Frege (1848–1925) devise a notational system capable of expressing the relevant logical relations. Virtually the whole of modern logic rests on Frege's work. This chapter and the next concern a Fregean language, $\mathcal{L}_p$, which provides a framework for representing and exploiting a range of logical relations absent from $\mathcal{L}_s$. We are not leaving $\mathcal{L}_s$ behind: $\mathcal{L}_s$ is simply absorbed into $\mathcal{L}_p$.

4.01 TERMS

The English sequence with which we opened the chapter expresses a valid argument. If you have doubts, try to think of circumstances under which it is true both that all people are mortal and that Socrates is a person, but false that Socrates is mortal. Although valid, the sequence's validity turns on the internal structure of the sentences that make it up, an internal structure that eludes the resources of $\mathcal{L}_s$.

Consider another English sequence

> All philosophers are clever.
> <u>Some Newfoundlanders are philosophers.</u>
> Therefore, some Newfoundlanders are clever.

This sequence, too, appears valid: we cannot imagine circumstances under which the premises are true and the conclusion is false. Suppose we replace nouns and adjectives in the sequence with letters and represent it schematically

> All P are Q
> <u>Some R are P</u>
> Therefore, some R are Q

We can substitute any nouns and adjectives we please for P, Q, and R, and the resulting sequence is valid. For instance

> All reporters are bigots.
> <u>Some redheads are reporters.</u>
> Therefore, some redheads are bigots.

This suggests that the validity of the sequence turns, not on its content, but on its structure. Further, the pertinent structural features are internal to the atomic sentences that make it up.

Consider a third sequence

> Some reporters are bigots.
> <u>Some redheads are reporters.</u>
> Therefore, some redheads are bigots.

This resembles the previous sequence with respect to the terms substituted for P, Q, and R.

> Some *P* are *Q*
> Some *R* are *P*
> Therefore, some *R* are *Q*

The second sequence differs from the first only in the substitution of "some" for "all" in its first sentence. Does this matter? The previous sequence was valid; this one is not. We can imagine circumstances under which the premises are true and the conclusion is false. Some reporters might be bigots and some redheads might be reporters without there being any bigoted redheaded reporters. So the truth of the premises is consistent with the falsehood of the conclusion.

Finally, consider the sequence

> All reporters are bigots.
> Some redheads are bigots.
> Therefore, some redheads are reporters.

This sequence returns to the"all"/"some" pattern of the original sequence, but the *P*, *Q*, *R* elements are differently distributed.

> All *P* are *Q*
> Some *R* are *Q*
> Therefore, some *R* are *P*

What is the effect on the sequence's validity? We can easily envisage circumstances under which reporters and some redheads are bigots even though no redheads are reporters. The sequence, then, is invalid.

The validity and invalidity of sequences like these evidently depends on (1) the arrangement of terms associated with *P*, *Q*, and *R*, and (2) the "all"/"some" pattern. Some arrangements of *P*s, *Q*s, *R*s, "all," and "some," result in a valid sequences, and others do not.

Let us focus on the *P*s, *Q*s, and *R*s. These clearly do not represent sentences; they represent *general terms*. General terms include common nouns (redhead, reporter), adjectives (red, wise), and intransitive verbs (leaps, sits). Sentences are true or false; general terms are *true of* (or fail to be true of) objects. "Redhead" holds true of each redhead, "red" is true of each red thing, and "leaps" is true of everything that leaps. This is not to say that a general term must be true of something. "Mermaid" and "phlogiston" are true of nothing.

Terms exhibit superficial grammatical differences. We say "Socrates is a philosopher," but "Socrates is wise," not "Socrates is *a* wise"; we say

"Socrates sits," but not "Socrates philosophers." There is less to these differences than meets the eye. In the interest of uniformity, we could paraphrase "Socrates is wise" as "Socrates is a wise thing" and "Socrates sits" as "Socrates is a sitting thing."

In addition to common nouns, adjectives, and intransitive verbs, general terms include ordinary transitive verbs—"sees," "admires," "lifts"—and comparative constructions having the form of complex transitive verbs—"is wiser than," "is taller than," "is between." These terms are true, not of objects considered singly, but of ordered pairs (or triples, or, more generally, of ordered n-tuples) of objects. An ordered pair of objects is a collection of two objects taken in a particular order. "Is taller than" is true of every ordered pair of objects whose first member is taller than its second. "Is between" is true of every ordered triple of objects the first member of which is between the remaining two members.

We can bring a measure of order to all this by regarding general terms as verbs, some of which are commonly expressed nonverbally. "Wise" and "red" are adjectives, but we can regard them as components of the complex intransitive verbs "is wise," and "is red." Returning to the list above, we find that "is redheaded," "is bigoted," "is a philosopher," "is a sitting thing," and so on provide serviceable alternatives to the originals.

Not every general term has a single word as an English equivalent. "Is a redheaded reporter" is, from the point of view of English, a complex general term true of redheaded reporters; "is a black bowlegged swan" is a general term true of black bowlegged swans.

The class or set of objects of which a general term is true is the *extension* of the term. The extension of the term "is a redheaded reporter" is the set of redheaded reporters; the extension of the term "is a mermaid" is the empty set. "Is a mermaid," "is a griffin," and "is a square circle" thus have a common extension. These terms are true of nothing at all.

General terms are to be distinguished from *singular terms*. A singular term purports to designate a unique object. Singular terms include proper names like "Socrates," "Ohio," and "Athena." They include, as well, descriptions, at least those intended to designate unique objects, *definite descriptions*: "the teacher of Plato," "the state in which Cleveland is located," "the Greek goddess of wisdom." Combining a general term and a singular term yields a *predication*, a simple sentence in which a general term purports to be true of an object (or an ordered n-tuple of objects); the English sentences below express simple predications:

> Socrates is a philosopher.
> The Greek goddess of wisdom leaps.

Socrates admires Athena.
Ohio is rectangular.

Although we count both proper names ("Socrates," "Ohio") and definite descriptions ("the teacher of Plato," "the state in which Cleveland is located") as singular terms, they differ otherwise. How they differ is a topic better approached after a consideration of terms in L_p.

───── **Exercises 4.00** ─────

Circle each singular term and draw a box around each general term in the sentences below. Indicate with a number (1, 2, 3, …) whether a general term is true of objects taken singly, ordered pairs of objects, ordered triples, or….

1. Callie is tall.

2. Joe is taller than Iola.

3. Callie is taller than Joe.

4. If Callie is taller than Joe and Joe is taller than Iola, then Callie is taller than Iola.

5. Callie is taller than Joe or Iola.

6. Gertrude sits between Frank and Joe.

7. Callie admires Fenton.

8. Fenton admires himself.

9. Chet dislikes Fenton only if Fenton admires himself.

10. Iola is shorter than Callie or Joe, but taller than Fenton.

11. Fenton gives The Sleuth to Frank and Joe.

12. Callie and Iola live in Bayport.

13. Fenton is a detective but Gertrude isn't.

14. If Fenton is a detective, he admires Frank and Joe.

15. Frank and Joe are brothers.

4.02 TERMS IN L_p

Formal languages like L_s and L_p provide a perspective on language that can illuminate natural languages like English. Our coming to appreciate the role of terms in L_p will enable us to see more clearly their role in English, which is often disguised or hidden in thickets of grammatical complexity. This section and the next concern the elements of L_p.

Let us designate the uppercase letters, $A–Z$, as *predicates* or *predicate constants*. These function in L_p not as sentences, but as general terms. We can represent the general term "is wise" by the predicate W, the general term "is a philosopher" by the predicate P, and so on. The lowercase letters, $a–t$, the *individual constants*, function as proper names do in English. We might use the individual constant s as the name "Socrates," the individual constant o as the name "Ohio," and the individual constant m as the name "Mars." The lowercase letters $u–z$ are *individual variables*. Variables have a distinctive role that will be discussed presently.

The simplest sentences in L_p are *predications*: individual constants paired with predicates. Consider the English predication

Socrates is a philosopher.

Suppose we let s stand in for the name "Socrates," and P express the general term "is a philosopher." Then we can represent this sentence in L_p as follows:

$$Ps$$

A general term like "is a philosopher" appears in L_p as a predicate true of, or not true of, individual objects. The term is true of the object named by "Socrates," and not true of the object named by "Napoleon." Predicates are introduced in the metalanguage by means of predicate letters and variables (lowercase letters, $u–z$). In the sentence above, we interpret Px as "x is a philosopher." If we interpret s as "Socrates" and a as "Athena," then

$$Ps$$

expresses in $\mathcal{L}_p$ what the English sentence

Socrates is a philosopher

expresses, and

$$Pa$$

expresses in $\mathcal{L}_p$ what is expressed by the English sentence

Athena is a philosopher.

Think of general terms in English—"is wise," "is a philosopher"—as incomplete symbols with slots that must be filled to yield a self-standing symbol. These slots are called *argument places*, and items filling the slots, *arguments*. The terms "is wise" and "is a philosopher" each have a single argument place. They correspond to one-place predicates in $\mathcal{L}_p$. A one-place predicate, a predicate with a single argument place, is called a *monadic predicate*. Px, then, is a monadic predicate. Here the variable x serves as a metalinguistic placeholder, a dummy symbol occupying the argument place in much the way the "—" in "—is a philosopher" does. We can turn Px into a sentence in the object language by replacing the x with an individual constant: Ps, Pa, and so on.

Now consider the English sentence

Socrates admires Athena.

The sentence includes two proper names, "Socrates" and "Athena," and a general term, "admires." "Admire," in English is a transitive verb expressing a general term true, not of objects taken singly, but of ordered pairs of objects. This is reflected in $\mathcal{L}_p$ by attaching to the predicate, not one, but two singular terms

$$Asa$$

Here, the $\mathcal{L}_p$ predicate corresponding to "admires" is a two-place predicate, a predicate with two slots or argument places corresponding to the two —s in "—admires—." A predicate of this sort is called a *dyadic* or *two-place predicate*. We introduce two-place predicates into $\mathcal{L}_p$ in a special way. We might say, for instance, that $A①②$ = "① admires ②." This means that given our previous interpretations of s and a, if we place s in the ① slot and a in the ② slot, the result is a predication true just in case Socrates admires Athena. Conversely, if we put a in the ① slot and replace ② with s, the result is a sentence true if and only if Athena admires Socrates.

General terms, then, can be monadic ("is wise," "is bigoted") and dyadic ("admires," "is taller than"). A monadic general term is true of objects; a dyadic term is true of ordered pairs of objects. Such terms are represented in $\mathcal{L}_p$ by one- and two-place predicates, respectively. Some terms are true (or not) of ordered *triples* of objects. The English general term "is between" can be represented in $\mathcal{L}_p$ by a three-place, *triadic* predicate: $B①②③$ = "① is between ② and ③." Given this interpretation, we can translate the English sentence

Clio is between Euterpe and Melpomene

into $\mathcal{L}_p$ as

Bcem

Once an interpretation is specified for $B①②③$, and once it is settled that c = "Clio," e = "Euterpe," and m = "Melpomene," we cannot change the order of the individual constants without changing the truth conditions of the sentence. The sentence

Bmce

is true just in case the English sentence

Melpomene is between Clio and Euterpe

is true.

Monadic, dyadic, and triadic predicates in $\mathcal{L}_p$ correspond to monadic, dyadic, and triadic general terms in English. How far can we go in this direction? We are assuming that sentences are finite in length. This limits to a finite number the number of argument places a predicate can have. We are left with the possibility of predicates and corresponding general terms, with any finite number of argument places. Had we the patience, we might concoct a predicate true of, and only of, ordered sets containing one hundred objects, the United States Senate, for instance. Let us use the phrase "n-place predicate" as a way of designating a predicate while leaving open whether it is monadic, dyadic, triadic, or something more.

Words and World

What do singular terms and general terms designate or "correspond to" in the world?

One traditional answer is that singular terms correspond to particulars, and general terms correspond to properties and relations. Thus the singular term "Socrates" designates the particular, Socrates, the general term "is wise" designates the property had by things that are wise, and "is braver than" designates the relation one thing has to another when it is braver than the other.

The distinction between particulars, on the one hand, and properties and relations, on the other hand, has struck many philosophers as fundamental. Particulars are said to be unique, dated individuals. Properties, in contrast, are repeatable; they can be shared (or "instantiated") by distinct particulars.

Do properties exist, as Plato thought, independently of particulars? Are there bare, propertyless particulars? Or are particulars bundles of properties?

In characterizing L_p, we can remain neutral on such questions. The result will be a language in which the world can be described in different ways—as including or lacking Platonic properties, as including or lacking propertyless particulars.

What L_p will provide is a way of regimenting talk about the world so that it is clear what sorts of entity we are committing ourselves to when we say what we do. (See § 4.15 below.)

Armed with a supply of predicates and individual constants, we can represent in L_p simple English sentences, what we earlier dubbed predications. If we reintroduce our old friends the connectives ($\neg$, $\wedge$, $\vee$, $\supset$, $\equiv$), we can dramatically broaden the scope of translation to include truth functions of these simple sentences. Take the sentence

If Euterpe is wise, then she is wiser than Clio.

Supposing Wx = "x is wise," and $W①②$ = "① is wiser than ②," we can translate the sentence into L_p as follows:

$$We \supset Wec$$

What of the English sentence

If Euterpe is wiser than Clio and Melpomene,
then she is wiser than Terpsichore.

Translation into L_p requires only that we construct appropriate predications and combine these with connectives just as we did in L_s

$$(Wec \land Wem) \supset Wet$$

The English sentence

If neither Clio nor Melpomene is wiser than Euterpe,
then Terpsichore is not wiser than Euterpe.

can be translated into L_p as

$$\neg(Wce \lor Wme) \supset \neg Wte$$

The technique for translating English sentences into L_s extends smoothly to the representation of simple predications and truth functions of predications in L_p. Matters become more complicated—and more interesting—when we move beyond predications.

---Exercises 4.01---

Construct L_p equivalents for the sentences in the previous exercise (4.00).

4.03 QUANTIFIERS AND VARIABLES

Predications in L_p resemble atomic sentences in L_s. Were L_p limited to predications and truth functions of predications, it would be logically on a par with L_s. The power of L_p arises, not from its capacity to express predications, but from its capacity to represent predications in a way that expresses *generality* of the sort exhibited by the syllogism with which we opened the chapter.

We find it useful to use names to designate individuals and properties that are important to us: Socrates, the Milky Way, water, oxygen. Most objects lack names. We identify them *ostensively*—by exhibiting them, or pointing to them—or by means of *descriptions*. I might, on a whim, name the desk in my office, but I am more likely to identify it simply as *that desk over there* or *the desk in my office*. This is unsurprising. Were we obliged always to refer to objects by name, communication would falter. Were I to decide to call my desk *Clyde*, you would be at a loss to understand my request to retrieve an object left on Clyde. You would have no way of knowing that I was referring to my desk unless you knew its name in advance. The world contains too many things for us to designate them all by names: a language that relied exclusively on names to designate objects would be unlearnable by finite creatures like us.

Let us focus on singular terms generally. Singular terms include names and descriptions, linguistic devices purporting to designate unique objects. Thus far we have concentrated exclusively on predications featuring proper names. In moving beyond names, the real power of L_p is revealed. Much of our talk about the world, and virtually all of our scientific talk, is framed in general terms. Consider, for instance the sentences

> The man in the corner is a spy.
> Whales are mammals.
> All people are mortal.
> Some planets have more than one moon.

Each of these sentences comments on the world. None does so by means of a name. L_p is capable of expressing such sentences in a way that makes their logical features transparent.

Why is generality important? You might attempt to express generality in a language by *listing* individuals. You might, for instance, try to paraphrase the English sentence

> Whales are mammals

by first giving a name to each whale, and then saying of each object thus named that it is both a whale and a mammal. The technique, however, besides being impossibly cumbersome, would neglect an essential feature of the original sentence. In referring to the class of whales, we take the class to be *open-ended.* Members may be added to it or subtracted from it indefinitely. In allowing for reference to unnamed individuals and for open-endedness, generality differs essentially from lists, even exhaustive lists, of individuals.

Names

In $\mathcal{L}_p$, lowercase letters *a–t* function as names. The role of these letters is comparable to the role of *proper names* in a natural language: names are used to designate or refer to objects.

What of names used to designate more than one object? And what of names whose objects are nonexistent?

We use the sign "Socrates" to designate the Greek philosopher Socrates, the teacher of Plato. But many people have been called Socrates. What connects our use of this sign to one of these people?

There is nothing in a sign itself that attaches it to one object rather than another. In *using a sign as a name we* fix its reference to a particular individual. The same sign can be used by different people on different occasions to designate different individuals.

What of names that lack bearers? Names purport to name objects, but can fail to do so. The existence of a name in a language does not guarantee the existence of an object corresponding to the name. A sign counts as a name when its job is to designate a unique object. If the object does not exist, the sign fails to perform its job, but through no fault of its own.

When we use a sign as a name, we commit ourselves to the existence of an individual corresponding to the name. There may, nevertheless, fail to be such an object.

Generality is introduced into L_p by means of *quantifiers* and *variables*. The lowercase letters *a* through *t* are used in L_p as names. The remaining lowercase letters, *u*, *v*, *w*, *x*, *y*, and *z*, function as variables ranging over objects. The idea is familiar from mathematics. We use 1, 2, 3, 4 as names of numbers. We can use *x*s and *y*s as variables ranging over numbers. We can say $x^2 + y^2 = z^2$. We understand the *x*s and *y*s here to stand for numbers, just not for any particular numbers.

Variables have an analogous role in L_p. Suppose that we represent the English sentence "Socrates is wise" as follows:

$$Ws$$

Then the expression

$$Wx$$

might be interpreted as ascribing wisdom to an arbitrary individual, *x*.

In fact, as we noted in § 4.02, the expression *Wx* is not a proper sentence of L_p. It resembles an English expression of the form

—is wise.

This expression is not an English sentence, although it could easily be turned into a sentence in one of two ways. We could replace "—" with a *name*

Athena is wise

we could replace the "—" with a *description*

The goddess of wisdom is wise

or we could replace the "—" with the English counterpart of a *quantifier*

Someone is wise.
Everyone is wise.

Variables in L_p cannot stand on their own. They are designed to be used together with appropriate quantifiers.

L_p features two quantifiers:

Universal quantifier: $(\forall \alpha)$
Existential quantifier: $(\exists \alpha)$

Each quantifier consists of a symbol—an inverted *A* or an inverted *E*—together with a variable. In the examples above, the Greek letter α (alpha) is used as a metalinguistic variable ranging over individual variables in L_p. (We are obliged to resort to Greek letters because L_p itself exhausts the Roman alphabet.) In putting quantifiers to use in sentences, the αs above would be replaced by some individual variable, *u–z*: $(\forall x)$, $(\forall y)$, $(\exists x)$, $(\exists y)$, and so on.

     Universal quantifiers express what we express in English by the phrases

all α

every α

any α

Existential quantifiers can be read as

some α

at least one α

Quantifiers in L_p, like their natural language counterparts, are never used in isolation. A quantifier is always attached to some other expression.

     The simplest quantified sentences are made up of a single quantifier and a single monadic predicate

$$(\forall x)\,Wx$$

which, using Wx to mean "x is wise," can be read

For all x, x is wise (i.e., everything is wise).

An existentially quantified counterpart

$$(\exists x)\,Wx$$

could be read as

There is at least one x, such that x is wise
(i.e., something is wise).

The metalanguage variable α has been replaced by an authentic L_p variable, x, in each quantifier occurrence. An x has been appended to the predicate W so that the variable occurring in the quantifier *matches* the variable occurring in the argument place of the predicate. This insures that the quantifier "picks up" the variable.

Pidgin $\mathcal{L}_p$

In moving between English and $\mathcal{L}_p$, it will sometimes be useful to employ "pidgin $\mathcal{L}_p$," a mixture of English and $\mathcal{L}_p$. We can read quantifiers in pidgin $\mathcal{L}_p$ as

$(\forall x)$: For all x....
$(\exists x)$: There is an x....

Complex expressions and sentences can be read in pidgin $\mathcal{L}_p$ as well. The standard forms of quantified sentences, for instance, can be read in pidgin $\mathcal{L}_p$ as follows:

$(\forall x)(Px \supset Qx)$: For all x, if x is P, then x is Q.
$(\exists x)(Px \wedge Qx)$: There is an x such that x is P and x is Q.

Think of variables occurring with predicates as *pronouns*, grammatical elements whose significance is determined by relations they bear to other elements in the sentence. The significance of a variable/pronoun is fixed by the quantifier with which it is associated and by the predicate (or predicates) whose argument place (or places) it fills. This aspect of variables will become clearer as we take up more sentences.

Differences in the logical structure of universally and existentially quantified expressions surface when we consider more complex sentences. One difference can be illustrated by means of examples. Consider the English sentence

All plants are green.

Letting Px = "x is a plant" and Gx = "x is green," we can translate this sentence into $\mathcal{L}_p$ as follows:

$$(\forall x)(Px \supset Gx)$$

In pidgin $\mathcal{L}_p$

For all x, if x is a plant, then x is green.

Note the occurrence of the variable x both in the quantifier and in the argument places of the predicate letters. The quantifier ties together these occurrences of x in a way that is brought out nicely by the pidgin $\mathcal{L}_p$ gloss accompanying the sentence.

Compare this universally quantified sentence with an existentially quantified counterpart

Some plants are green.

The sentence can be translated into $\mathcal{L}_p$ as

$$(\exists x)(Px \wedge Gx)$$

In pidgin $\mathcal{L}_p$

There is at least one x such that x is a plant and x is green

Think of these sentences as *paradigms*, typifying examples of universally and existentially quantified sentences. We see immediately that universally quantified sentences are treated as conditionals; existentially quantified sentences are taken to be conjunctions. In this regard, $\mathcal{L}_p$ illuminates the logic of sentences featuring "all" and "some," a logic disguised in the English originals.

Consider again the English sentence

All plants are green.

Predicates function as general terms. A general term is true of, or not true of, objects. (Or, if it is not monadic, it is true of, or not true of, ordered n-tuples of objects; we can omit this qualification here.) We characterized the class or set of objects of which a general term is true as the *extension* of the term. A term's extension consists of the set of every object *satisfying* the term. We can interpret the sentence above as saying something about the relation classes of objects bear to one another.

If anything is in the class of plants,
then it is in the class of green things.

The class relation expressed by the sentence can be depicted by means of a diagram:

The class of plants is *included* in the class of green things. The *inclusion* relation is nicely captured by a universally quantified conditional sentence that says

> Take any object at all. Call that object x. If x is in the class
> of plants, then x is in the class of green things.

Does it follow that if something is green it is a plant? No. The sentence leaves open that the class of green things has members that are not plants.

Compare this with the sentence

> Some plants are green.

This sentence might be taken to mean

> There is at least one thing that is both a member of the
> class of plants and a member of the class of green things.

Diagrammatically represented:

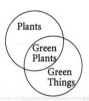

Universally quantified sentences express the class inclusion relation; existentially quantified sentences express class *intersection*. They say that the intersection of the class of plants and the class of green things is not empty; at least one object is a member of *both*.

These simple relationships—class inclusion and class intersection—are carried through the use of universal and existential quantifiers even in complex sentences. With few exceptions, universally quantified sentences and sentence parts are built around a conditional core; existentially quantified sentences and sentence parts are built around conjunctions.

Exercises 4.02

Translate the English sentences below into $\mathcal{L}_p$. Let Sx = "x is a sleuth"; Ax = "x is an aunt"; Cx = "x is cautious." Use appropriate lowercase letters for names.

1. Frank and Joe are sleuths.

2. Gertrude isn't a sleuth; she's an aunt.

3. Some aunts are sleuths.

4. If some aunts are sleuths, then some sleuths are aunts.

5. If Gertrude is a sleuth, then some aunts are sleuths.

6. Fenton is both cautious and a sleuth.

7. All sleuths are cautious.

8. Every aunt is cautious.

9. Iola is neither an aunt nor cautious.

10. If all aunts are cautious, then Gertrude is cautious.

11. Gertrude is a sleuth only if some aunts are sleuths.

12. Some sleuths are both cautious and aunts.

13. Frank and Joe are cautious if and only if every sleuth is cautious.

14. Every aunt, if she is cautious, is a sleuth.

15. If Gertrude is a sleuth, then at least one aunt is a sleuth.

4.04 BOUND AND FREE VARIABLES

In $\mathcal{L}_p$, every quantifier incorporates a variable that ties together the expression to which it is appended. Variables occurring in the argument places of predicates resemble pronouns in English, terms that share a designation within a sentence. The mechanism in $\mathcal{L}_p$ is straightforward: a quantifier picks up a variable in a predicate when the variable in the quantifier *matches* the variable in the predicate. We read the $\mathcal{L}_p$ sentence

$$(\forall x)(Fx \supset Gx)$$

as

All Fs are Gs.

This reading is made possible by the matching pattern of variables. "All" refers to Fs and Gs because the variable contained in the universal quantifier, x, matches the variable filling the argument places in the predicate expressions. Further, the matching variables insure that the objects said to be Fs are the same as those said to be Gs.

A quantifier *picks up* any matching variable falling within its *scope*. The scope of a quantifier is to be understood exactly as we understand the scope of a negation sign. It includes everything in the expression to its immediate right. Consider the $\mathcal{L}_p$ sentences below:

$$\neg Fa \supset Ga$$
$$\neg(Fa \supset Ga)$$

In the first sentence, the scope of the negation sign includes only Fa. The scope of the negation sign in the second sentence includes everything falling within the matching parentheses to its right, $Fa \supset Ga$.

In exactly the same way, the scope of the universal quantifier, $(\forall x)$, in the two expressions below includes the expression to its immediate right:

$$(\forall x)Fx \supset Gx$$
$$(\forall x)(Fx \supset Gx)$$

In the first case, only Fx falls within the scope of the quantifier. In the second case, the entire expression within parentheses, $Fx \supset Gx$, falls within its scope.

Let us say that a variable is *bound* just in case (1) the variable falls within the scope of a quantifier, and (2) the variable matches the variable contained in the quantifier. If a variable is not bound, it is *free*. (Occurrences of variables inside quantifiers are neither bound nor free.) In the first example above, the *x* in *Fx* is bound by the universal quantifier to its immediate left, but the *x* in *Gx* is free. In both cases the variable *x* matches the variable in the universal quantifier, but only the *x* in *Fx* falls within its scope.

No sentence of $\mathcal{L}_p$ can contain free variables. We can insure that variables in sentences containing quantifiers function appropriately as pronouns only if those variables are picked up by quantifiers. The occurrence of a free variable in an $\mathcal{L}_p$ expression is comparable to the occurrence of "—" in the midst of what otherwise would be an ordinary English sentence. Suppose *Fx* = "*x* is friendly" and *Gx* = "*x* is good." Then the first expression above might be represented in English as

If everything is friendly, then — is good.

This is not quite an English sentence.

Exercises 4.03

Examine the expressions below and indicate (1) the scope of each quantifier, and (2) which variables, if any, occur freely. Indicate quantifier scope by means of a line, and circle free variables.

Example: (∀x)(Fx ⊃ Gx) ∧ Hⓧ

1. (∀x)Fy ⊃ Gx

2. (∃x)(Fx ∧ Gx)

3. (∀x)(Fx ⊃ Gx) ⊃ (∃x)(Fx ∧ Gx)

4. (∃y)(Fx ∧ Gx) ∧ (∀x)(Fy ⊃ Gy)

5. (∀x)((Fx ⊃ Gx) ⊃ (Hx ∧ Ix))

6. (∃y)((Fy ∧ Gy) ∧ (Hy ∧ Iy))

7. ((∃y)(Fy ∧ Gy) ∧ (Hy ∧ Iy))

8. (((∃y)Fy ∧ Gy) ∧ (Hy ∧ Iy))

9. $Fa \supset Ga$

10. $(\forall x)(Fx \supset Gx) \supset Ha$

4.05 NEGATION

The occurrence of negation in quantified sentences poses no special problems. Quantifiers can fall within the scope of negation signs, and *vice versa.* Compare the English sentences

Not all sleuths are married.
All sleuths are unmarried.

Both sentences are variants of the core sentence

All sleuths are married.

Letting Sx = "x is a sleuth," and Mx = "x is married," we can translate this core sentence into L_p as

$$(\forall x)(Sx \supset Mx)$$

How do we translate "Not all sleuths are married" into L_p? The sentence is the negation of the original—"Not all...."—so we obtain

$$\neg(\forall x)(Sx \supset Mx)$$

What about "All sleuths are unmarried?" In this case we negate the predicate expression, "x is married," Mx, not the quantifier

$$(\forall x)(Sx \supset \neg Mx)$$

If you are like most people, you will find it helpful to turn to pidgin L_p both in the course of translating English sentences into L_p and in figuring out the meanings of L_p sentences. The sentence above, in pidgin L_p is

For all x, if x is a sleuth, then
it's not the case that x is married.

The same principles apply to existentially quantified sentences. Let

$$(\exists x)(Sx \wedge Mx)$$

represent the English sentence

Some sleuths are married.

How would we translate the English sentences below into $\mathcal{L}_p$?

No sleuths are married.
Some sleuths are unmarried.

The first sentence tells us that it is not the case that at least one sleuth is married

$$\neg(\exists x)(Sx \wedge Mx)$$

(Remember: we are taking the existential quantifier to mean "some" in the sense of "at least one.") If it is not the case that at least one sleuth is married, then no sleuths are married. If some sleuths are unmarried, then something is such that it *is* a sleuth and *is not* married

$$(\exists x)(Sx \wedge \neg Mx)$$

In pidgin $\mathcal{L}_p$

There is an x such that x is a sleuth
and it's not the case that x is married.

In mastering quantified sentences and sentence parts, we should not lose sight of sentences containing names. Consider the sentence

Some Greek is a philosopher.

Letting Gx = "x is a Greek" and Px = "x is a philosopher," we can translate this sentence into $\mathcal{L}_p$ as

$$(\exists x)(Gx \wedge Px)$$

What about the sentence

If Socrates is a philosopher,
then some Greek is a philosopher.

The antecedent of this conditional contains a name, an individual constant,
not a variable

$$Ps \supset (\exists x)(Gx \wedge Px)$$

Another example:

If Socrates is feared by some philosopher,
he is feared by every philosopher.

Letting $F①② =$ "① fears ②," we can translate the sentence into L_p as

$$(\exists x)(Px \wedge Fxs) \supset (\forall x)(Px \supset Fxs)$$

Translating this into pidgin L_p:

If there is an x such that x is a philosopher and x fears
Socrates, then for all x, if x is a philosopher, x fears Socrates.

Variables take on significance only when matched with a quantifier, but
individual constants stand on their own.

Exercises 4.04

Translate the English sentences below into L_p. Let $Sx =$
"x is a sleuth"; $Wx =$ "x is wily"; $Kx =$ "x is kidnapped." Use
appropriate lowercase letters for names.

1. Every sleuth is wily.

2. If Chet is kidnapped, then some sleuths aren't
   wily.

3. If some sleuths aren't wily, then not all sleuths
   are wily.

4. If all sleuths are wily, then if Fenton isn't wily, he
   isn't a sleuth.

5. If Gertrude and Callie are kidnapped, then no sleuth is wily.

6. No sleuth fails to be wily.

7. Not every sleuth fails to be wily.

8. If some sleuths are kidnapped, then not all sleuths are wily.

9. If some sleuths aren't kidnapped, then not all sleuths fail to be wily.

10. If neither Gertrude nor Callie is kidnapped, then Iola is wily.

11. If no sleuths are wily, then some sleuth is kidnapped.

12. Some sleuths are kidnapped even though every sleuth is wily.

13. If not all sleuths are kidnapped, then Frank or Joe isn't kidnapped.

14. Gertrude is kidnapped only if some sleuth isn't wily.

15. If a sleuth is wily, he isn't kidnapped.

4.06 COMPLEX TERMS

We have equated predicates in $\mathcal{L}_p$ with general terms in English. Objects in the extension of a general term are members of a set or class of objects, each of which is such that the general term is true of it. Suppose the predicate S is interpreted as expressing the general term "is a swan" and the predicate B is taken to express "is black." Then the extension of S is the class of swans and the extension of B is the class of black objects.

English includes a plentiful supply of words expressing general terms. When the need arises, we make up new words and add them to our language. Meteorologists have an expanded vocabulary for the description of clouds, entymologists for the description of bugs, physicists for the description of particles. Words can fall into disuse when we lose interest in the distinctions they mark. It is reported that eighteenth century English contained words for emotions lost to twentieth century English. The distinctions remain, though our interest in making them has diminished.

Suppose general terms were expressible only as individual words, and singular terms were limited to names. Language use would be a dreary affair, confined to situations connected to the immediate surroundings of speakers and hearers. As it is, we find it possible to speak of absent and unnamed objects by using descriptions: "the desk in my office," "the magnolia tree in my front yard," "the goddess of wisdom." Another device is available for the manufacture of general terms. English contains the general terms "is a swan" and "is a black thing." Putting these together we can manufacture a new general term, "is a black swan." The extension of this term includes objects belonging to the *intersection* of the class of swans and the class of black things. It includes everything that is *both* a swan and black:

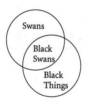

Similarly, we can manufacture a new general term, "is a black bowlegged swan." The term's extension is the intersection of the class of black things, the class of swans, and the class of bowlegged things. The technique can be extended *ad lib*.

Let us call terms manufactured in this way *complex terms*. Sentences containing complex terms are easily translatable into $\mathcal{L}_p$. Consider, for instance, the sentence

Every black swan is graceful.

We treat "black swan" as a complex term, and represent the sentence as follows (assuming that $Sx =$ "x is a swan," $Bx =$ "x is black," and $Gx =$ "x is graceful"):

$$(\forall x)((Sx \land Bx) \supset Gx)$$

In pidgin $\mathcal{L}_p$

For all x, if x is a swan and x is black (i.e., if x is a member of both the class of swans and the class of black things), then x is graceful.

The sentence follows the paradigm for universally quantified expressions illustrated by

All swans are graceful.

In $\mathcal{L}_p$

$$(\forall x)(Sx \supset Gx)$$

The antecedents of the conditionals differ—the first sentence features a complex term in the form of a conjunction in its antecedent, and the second does not—but the pattern is the same. In each case, one class is said to be included in another.

What of existentially quantified expressions containing complex terms? Consider the English sentence below:

Some black swans are graceful.

This can be translated into $\mathcal{L}_p$ as

$$(\exists x)((Sx \wedge Bx) \wedge Gx)$$

In both the universally quantified sentence above and in this sentence, $Sx \wedge Bx$ expresses the complex general term "is a black swan." The term is one for which we have no single word, so we must make do with a term cooked up on the spot. Ignoring that complication, the sentences exhibit the standard forms of universally and existentially quantified sentences, respectively.

Exercises 4.05

Translate the English sentences below into $\mathcal{L}_p$. Let $Sx =$ "x is a sleuth"; $Cx =$ "x is clever"; $C①② =$ "① is more clever than ②"; $Ex = $ x escapes. Use appropriate lowercase letters for names.

1. Every clever sleuth escapes.

2. Some sleuth is more clever than Fenton.

3. No sleuth is more clever than Gertrude.

4. Every sleuth is more clever than Chet.

5. If Fenton escapes, then some clever sleuth escapes.

6. If Gertrude doesn't escape, then no clever sleuth escapes.

7. If neither Gertrude nor Fenton escapes, then some clever sleuth doesn't escape.

8. Not every clever sleuth escapes.

9. Every clever sleuth fails to escape.

10. If some clever sleuth escapes, then some clever sleuth is more clever than Gertrude.

11. If Frank and Joe escape, then Gertrude is more clever than Callie or Iola.

12. Frank escapes only if some clever sleuth is more clever than Gertrude.

13. Not every clever sleuth fails to escape.

14. If some clever sleuth escapes, then every clever sleuth escapes.

15. Not every clever sleuth is more clever than Gertrude.

4.07 MIXED QUANTIFICATION

The translation of many English sentences into $\mathcal{L}_p$ requires more than one quantifier. When a sentence contains multiple quantifiers, the quantifiers and the variables they bind must be kept properly sorted out. Consider the English sentence

Cats like fish.

Taken out of context, this sentence is *quadruply* ambiguous. It could be used to mean any of the following:

All cats like all fish.
All cats like some fish.
Some cats like all fish.
Some cats like some fish.

Let us tackle each of these sentences, supposing that Cx = "x is a cat," Fx = "x is a fish," and $L①②$ = "① likes ②."
    First

All cats like all fish.

This sentence can be translated into $\mathcal{L}_p$ as

$$(\forall x)(Cx \supset (\forall y)(Fy \supset Lxy))$$

In pidgin $\mathcal{L}_p$

For all x, if x is a cat, then for all y,
if y is a fish, x likes y.

Four features of the translation deserve mention:

1. Every variable is bound by a quantifier. All occurrences of the variable x fall within the scope of the $(\forall x)$ quantifier, and all occurrences of the variable y fall within the scope of the $(\forall y)$.

2. The second quantifier, $(\forall y)$, falls within the scope of the first, $(\forall x)$. When this happens, each quantifier must incorporate a *distinct* variable. Otherwise the sentence would be ambiguous.

3. The connectives associated with each quantifier follow the paradigm. Universally quantified expressions typically feature conditionals, and so it is here. The first quantifier, $(\forall x)$, quantifies a conditional sentence that happens to have a conditional consequent; the second quantifier, $(\forall y)$, quantifies that conditional consequent.

4. The variables, x and y, function as pronouns. The pattern of variables, quantifiers, and predicate letters establishes that xs are cats, ys are fish. In the context of the sentence, the two-place predicate, Lxy, means that *cats* like *fish*. Had we written instead, Lyx, the liking relation would be reversed: because xs are cats and ys are fish, the sentence would say that *fish* like *cats*.

Let us look at $\mathcal{L}_p$ translations of the remaining sentences in the original list. In each case, the four points discussed above—suitably amended—apply.

All cats like some fish.

In $\mathcal{L}_p$

$$(\forall x)(Cx \supset (\exists y)(Fy \wedge Lxy))$$

And in pidgin $\mathcal{L}_p$

For all x, if x is a cat, then there is a y,
such that y is a fish and x likes y.

Here an *existential* quantifier, $(\exists y)$, appears in the consequent of a universally quantified conditional (reflecting its reference to *some* fish), and the connective is adjusted accordingly. As we should expect $\supset$ is paired with a universal quantifier, $(\forall x)$, and $\wedge$ with an existential quantifier.

An analogous pattern can be observed in translating

Some cats like all fish.

In $\mathcal{L}_p$

$$(\exists x)(Cx \wedge (\forall y)(Fy \supset Lxy))$$

And in pidgin $\mathcal{L}_p$

> There is an x, such that x is a cat,
> and for all y, if y is a fish, then x likes y.

Again, the pattern of quantifiers and connectives honors the paradigms. The sentence is an existentially quantified conjunction, the second conjunct of which is a universally quantified conditional.

Finally

> Some cats like some fish.

In $\mathcal{L}_p$

$$(\exists x)(Cx \wedge (\exists y)(Fy \wedge Lxy))$$

In pidgin $\mathcal{L}_p$

> There is an x, such that x is a cat,
> and there is a y such that y is a fish, and x likes y.

The sentence consists of an existentially quantified conjunction embedded as a conjunct within an existentially quantified conjunction.

Attention to sentences featuring mixed quantifiers leads inevitably to an observation about the significance of quantifier *order* within sentences. Consider the English sentence

> Every sailor detests some port.

The sentence, like many (most? all?) sentences taken out of context, is ambiguous. It can be used to mean that some particular port—Sydney, say—is detested by every sailor. Alternatively, the sentence can be used to mean that every sailor detests some port or other, a port that might vary among sailors: some detest Sydney, others detest San Diego, still others detest Reykjavik. These different interpretations of the English original are reflected in the order in which quantifiers are introduced. Thus, letting $Sx =$ "x is a sailor," $Px =$ "x is a port," and $D①② =$"① detests ②," the $\mathcal{L}_p$ sentence

$$(\exists x)(Px \wedge (\forall y)(Sy \supset Dyx))$$

says, in pidgin $\mathcal{L}_p$

> There is an x such that x is a port, and,
> for all y if y is a sailor, y detests x.

In English

> There is some port detested by every sailor.

Compare this sentence with

$$(\forall x)(Sx \supset (\exists y)(Py \wedge Dxy))$$

In pidgin $\mathcal{L}_p$

> For all x, if x is a sailor, then
> there is a y such that y is a port and x detests y.

And in ordinary English

> Every sailor detests some port (i.e., some port or other).

The order of occurrence of the variables used in the argument places of the predicate D①② (① detests ②) is significant. Their order reflects the pronominal character of variables generally. In the first sentence, x is specified as a *port*, hence it is the *object* of the detesting; in the second sentence, x is taken to be a member of the class of sailors, hence x is the *subject* of the detesting relation.

You will have noticed that variables appearing in the argument places in sentences need not match the metalinguistic variables used in specifying an interpretation of the $\mathcal{L}_p$ predicate. We said that Sx = "x is a sailor" and Px = "x is a port." Variables so used are mere place holders. The interpretation tells us that *whatever* goes in the argument place occupied by x in Sx is being said to be a sailor. If we subsequently write "Sy," we are writing "y is a sailor"; if we write "Ss," we are writing "Socrates is a sailor."

Exercises 4.06

Translate the English sentences below into $\mathcal{L}_p$. Let Sx = "x is a sleuth"; Cx = "x is a criminal"; Mx = "x is married"; $A①②$ = "① arrests ②." Use appropriate lowercase letters for names.

1. Every sleuth arrests some criminal.

2. No sleuth arrests any criminal.

3. Every criminal is arrested by some sleuth.

4. Some sleuth arrests every criminal.

5. Some unmarried sleuth arrests every married criminal.

6. Fenton arrests every unmarried criminal.

7. If every married sleuth arrests some unmarried criminal, then some criminal is unmarried.

8. Not every sleuth arrests some criminal.

9. Every sleuth fails to arrest some criminal.

10. Not every married sleuth fails to arrest some unmarried criminal.

11. All married criminals are arrested by some sleuth.

12. Every sleuth fails to arrest some unmarried criminal.

13. No sleuth arrests every married criminal.

14. If Fenton arrests every criminal, then he arrests every married criminal.

15. If not every married criminal is arrested by some sleuth, then some criminals aren't arrested by any sleuth.

4.08 TRANSLATIONAL ODDS AND ENDS

Consider the sentence

Some dog is missing.

How might we translate this sentence into L_p? By now it is second nature. Letting $Dx =$ "x is a dog," and $Mx =$ "x is missing," we obtain

$$(\exists x)(Dx \wedge Mx)$$

That is, in pidgin L_p

There is an x such that x is a dog and x is missing.

Now consider a superficially similar sentence

Something is missing.

It is tempting to read this sentence in pidgin L_p as

There is an x such that x is a thing, and x is missing,

and so to translate it

$$(\exists x)(Tx \wedge Mx)$$

The temptation is to be resisted. Saying that "x is a thing" is not to predicate anything of x. It is simply to indicate something to which predicates can be applied. To say, in L_p, that something is missing, we need only

$$(\exists x)Mx$$

This says

There is an x (i.e., there is something) such that x is missing.

The point extends to sentences sporting mixed quantifiers. Consider the English sentence

Some sailor is taller than every landlubber.

Assuming that $Sx =$ "x is a sailor," $Lx =$ "x is a landlubber," and $T①② =$ "$①$ is taller than $②$," we can translate the sentence into L_p as

$$(\exists x)(Sx \wedge (\forall y)(Ly \supset Txy))$$

Now consider the sentence

> Something is taller than any landlubber.

This goes into L_p as

$$(\forall x)(Lx \supset (\exists y)Tyx)$$

In pidgin L_p

> For all x, if x is a landlubber, then
> there is a y such that y is taller than x.

Finally consider the English sentence

> Everything is taller than something.

The sentence leaves utterly open the *sorts* of thing being compared

$$(\forall x)(\exists y)Txy$$

In pidgin L_p

> For all x, there is some y such that x is taller than y.

English words sometimes appear to function as terms when they do not. There are, as well, cases in which terms are hidden or disguised. Consider the English sentence

> Everyone is taller than someone.

The word "everyone" here could be paraphrased "every *person*"; "someone" could be paraphrased "some (or at least one) *person*." "Everyone" and "someone" differ fundamentally from "everything" and "something," despite a superficial similarity. Letting $Px =$ "x is a person," the sentence above is translated into L_p as

$$(\forall x)(Px \supset (\exists y)(Py \wedge Txy))$$

Another potentially troublesome English construction is illustrated by the sentence

A computer is a machine.

Although, at first glance, the sentence seems to refer to some particular computer (*a* computer...), a moment's reflection reveals that it is intended to say something about computers generally, namely that if it is true of something that it is a computer, then it is true of that thing that it is a machine. So the sentence is equivalent to

All computers are machines.

And, given that $Cx =$ "x is a computer," and $Mx =$ "x is a machine," this sentence is translated into L_p as

$$(\forall x)(Cx \supset Mx)$$

As usual in the case of translation, the aim is to produce a sentence whose truth conditions approximate those of the original. This obliges us to look beyond the surface structure of sentences we are translating so as not to trip over syntactic forms that are misleading when taken out of context.

We expect "any," "all," and "every" to signal the presence of a universal quantifier. Often they do. In the sentence below, "any" picks out every member of the class of hamsters

Any hamster that bites is dangerous.

Assuming that $Hx =$ "x is a hamster," $Bx =$ "x bites," and $Dx =$ "x is dangerous," this sentence can be translated into L_p as

$$(\forall x)((Hx \wedge Bx) \supset Dx)$$

But consider the superficially similar sentence

If any hamster bites, then some hamster is dangerous.

This sentence does not mean "If all (or every) hamster bites...," but "If some (at least one) hamster bites...," and, in consequence, it must be translated into L_p as

$$(\exists x)(Hx \wedge Bx) \supset (\exists x)(Hx \wedge Dx)$$

In this sentence, xs have been used in both quantifiers. This is permissible because the quantifiers do not overlap in scope. We could have translated the sentence

$$(\exists x)(Hx \wedge Bx) \supset (\exists y)(Hy \wedge Dy)$$

The sentences are logically equivalent; either is acceptable.

It is hazardous to look for the occurrence of particular words in sentences and take these as infallible signs of logical structure. The same word in different sentential contexts can express different meanings. We learn words by learning to use them in sentences. Consider the English sentence

Euterpe is bolder than Clio or Melpomene.

At first glance the sentence appears to express a disjunction. But does it? Is the sentence below an acceptable paraphrase?

Euterpe is bolder than Clio or
Euterpe is bolder than Melpomene.

The original sentence says that Euterpe is bolder than *either*. Thus

Euterpe is bolder than Clio
and Euterpe is bolder than Melpomene.

Once this is recognized, translation into $\mathcal{L}_p$ is a breeze. Letting $B①② =$ "① is bolder than ②":

Bec $\wedge$ Bem

We can think of the meaning of a word as the contribution that word makes to the truth conditions of sentences in which it occurs. We can now see that a word can affect the truth conditions of sentences differently depending on its context.

Translation Applied

Consider the English sentence

> If all horses are animals, then all heads of horses are heads of animals.

How might we put this into $\mathcal{L}_p$? (Let Hx = "x is a horse"; Ax = "x is an animal"; and $H①②$ = "① is the head of ②.") The sentence is a conditional whose antecedent is

> If all horses are animals, then....

Thus

$$(\forall x)(Hx \supset Ax) \supset$$

The consequent is trickier.

> ...all heads of horses are heads of animals.

This goes into $\mathcal{L}_p$ as

$$(\forall x)((\exists y)(Hx \wedge Hyx) \supset (\exists y)(Ay \wedge Hxy))$$

In pidgin $\mathcal{L}_p$: For all x, if there is a y such that y is a horse and x is the head of y, then there is a y such that y is an animal and x is the head of y. The whole sentence in $\mathcal{L}_p$ is

$$(\forall x)(Hx \supset Ax) \supset$$
$$(\forall x)((\exists y)(Hx \wedge Hyx) \supset (\exists y)(Ay \wedge Hxy))$$

[See W. V. Quine, *Methods of Logic*, 3d ed. (New York: Holt, Rinehart and Winston, 1972), pp. 142–43.]

Exercises 4.07

Translate the English sentences below into $\mathcal{L}_p$. Let Sx = "x is a sleuth"; Px = "x is a person"; Wx = "x is wily"; $A①②$ = "① admires ②." Use appropriate lowercase letters for names.

1. A sleuth is a wily person.

2. Someone is admired by every sleuth.

3. Every sleuth admires something.

4. No sleuth is admired by everyone.

5. If Fenton is a wily sleuth, he is admired by someone.

6. If Fenton isn't a wily sleuth, he is admired by no one.

7. Callie admires a wily sleuth.

8. If Callie admires Fenton, she admires a wily sleuth.

9. If a sleuth isn't wily, then nothing is wily.

10. A wily sleuth admires nothing.

11. A person who fails to be wily is not admired by anyone.

12. Not every wily sleuth is admired by Fenton.

13. Every wily sleuth is admired by someone.

14. Some wily sleuth is admired by everyone.

15. A wily sleuth is a sleuth admired by everyone.

4.09 IDENTITY

Identity is a relation every object bears to itself and to no other object. The concept of identity is the concept of *self-sameness*. If x and y are identical individuals, then x and y are *one and the same* individual. This way of putting it has a faintly paradoxical ring. How can *two* things be *one and the same* thing? The paradox is only apparent. If x and y are identical, then there are not *two* things, x and y, but only a single thing twice designated. We require a concept of identity precisely because things can be variously named and described. The identity relation enables us to indicate that two terms designate one and the same individual: Mark Twain and Samuel Clemens are one and the same person.

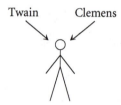

The identity relation in L_p holds among the objects of individual terms. If a and b are identical, the terms "a" and "b" designate one and the same individual. Cicero *is—is identical with*—Tully, the Morning Star *is* the Evening Star, Scott *is* the author of *Waverly*, the masked bandit *is* the dashing prince.

The "is" of identity must be distinguished from the "is" of predication. The sentences below exhibit the "is" of identity:

> The Morning Star is the Evening Star.
> Lewis Carroll is Charles Dodgson.

This "is" differs from the "is" of predication illustrated by these sentences:

> The Morning Star is bright.
> Carroll is English.

Here "is" appears as a component of a pair of general terms: "is bright," and "is English."

Two simple, though fallible, tests for the "is" of identity are worth mention. First, it is often possible to reverse the order of the terms flanking the "is" of identity without altering the truth conditions of the sentence in which it occurs. If the Morning Star is the Evening Star, then the Evening

Star is the Morning Star; if Carroll is Dodgson, then Dodgson is Carroll. This is usually not possible with the "is" of predication. It makes no sense to say "bright is the Morning Star" or "English is Carroll." Second, the "is" of identity can often be replaced by the phrase "is nothing but." The Morning Star is nothing but the Evening Star, and Carroll is nothing but Dodgson; but it is literally senseless to say that the Morning Star is nothing but bright, or that Carroll is nothing but English.

Identity is signalled in $\mathcal{L}_p$ by the identity sign, = (sometimes misleadingly called the equality sign). The sign can be used only flanked by names or variables

$$a = x$$
$$x = y$$

But *not*

$$Pa = Pb$$
$$(\exists x)Fx = (\exists x)Gx$$

The negation of an identity relation can be indicated either by means of a conventional negation sign or via a ≠. The sentences below are notational variants of one another:

$$\neg(a = b)$$
$$a \neq b$$

Both sentences deny that "a" and "b" designate one and the same individual. Either is acceptable.

We can take another notational liberty with the identity sign. Because the identity sign is always flanked by individual names or variables, we can omit parentheses without producing ambiguity. Instead of writing

$$(\forall x)(Px \supset (x = a))$$

we can write

$$(\forall x)(Px \supset x = a)$$

The identity sign locks together the x and a so that parentheses are redundant. The advantage of this convention will become evident later when we survey more complex sentences in which it is a challenge to keep track of iterated parentheses.

Identity

Everything is identical with itself and nothing else. It seems to follow that statements of identity must be either trivially true

Lewis Carroll = Lewis Carroll

or patently false

Lewis Carroll = Percy Bysshe Shelley.

But we know that

Lewis Carroll = Charles Dodgson

is an identity statement that is both true and informative. We require a concept of identity in part because names and objects are not perfectly correlated. We require an identity concept, as well, because it is possible to signify objects in many ways, and we need some way of indicating that objects variously signified are one and the same:

Lewis Carroll = the author of Jabberwocky,

the author of Jabberwocky =
the lecturer in mathematics.

We are now in a position to translate English sentences incorporating identity relations into $\mathcal{L}_p$. Consider the English sentence

> If Twain is an author and Clemens isn't,
> then Twain isn't Clemens.

Do not be confused by occurrences in the sentences of both the "is" of predication and the "is" of identity. The sentence can be translated into $\mathcal{L}_p$ without recourse to quantifiers (letting $Ax =$ "x is an author")

$$(At \land \neg Ac) \supset t \neq c$$

Consider a more complex sentence involving quantification:

Clemens is the only author.

Although it is not immediately obvious, this sentence harbors an occurrence of the identity relation. The sentence tells us both that Clemens is an author and that Clemens is the *only* author. How might this be expressed in $\mathcal{L}_p$? If Clemens is the only author, then anything that is an author, anything of which "is an author" is true, must be Clemens. Putting these together

$$Ac \wedge (\forall x)(Ax \supset x = c)$$

That is, in pidgin $\mathcal{L}_p$

Clemens is an author and, for all x,
if x is an author, then x is Clemens.

The sentence nicely illustrates the extent to which the identity relation can play a central but unobvious role in the logic of everyday talk. More generally: in looking carefully at the truth conditions of apparently unremarkable sentences, we may discover a formidable logical structure.

Exercises 4.08

Translate the English sentences below into $\mathcal{L}_p$. Let $Sx =$ "x is a sleuth"; $A①② =$ "① is the aunt of ②." Use appropriate lowercase letters for names.

1. Gertrude is Joe's Aunt.

2. Gertrude is Miss Hardy.

3. Gertrude is Joe's only aunt.

4. Joe's only aunt isn't a sleuth.

5. Frank's only aunt is Joe's only aunt.

6. Joe isn't Frank.

7. If Frank isn't Joe's aunt, then Frank isn't Joe's only aunt.

8. If Gertrude is Miss Hardy, then, if Gertrude is a sleuth, Miss Hardy is a sleuth.

9. If Gertrude is a sleuth and Miss Hardy isn't a sleuth, then Gertrude isn't Miss Hardy.

10. Either Miss Hardy is Joe's aunt or Miss Hardy isn't Gertrude.

11. If Miss Hardy is Joe's aunt, then Joe's aunt is a sleuth.

12. If Joe's aunt is a sleuth, Frank's aunt is a sleuth.

13. Frank's aunt is a sleuth only if she's Miss Hardy.

14. If Gertrude is Joe's aunt, she isn't a sleuth only if she isn't Miss Hardy.

15. Not every sleuth is Joe's aunt.

4.10 PUTTING IDENTITY TO WORK

In introducing L_p, we noted that singular terms encompass both names and descriptions. You might refer to a certain philosopher by means of a name, *Aristotle*, or by means of a *definite description*: "the teacher of Alexander." Definite descriptions are ubiquitous. A language that did not permit the construction of descriptions would be communicatively rigid. Mutual understanding would require the prior mastery of a stock of names. But how might these names be learned? Names are most often taught by means of descriptions, rather than ostensively, by exhibiting their bearers. We are taught that *Athena* is the Greek goddess of wisdom, that *Augusta* is the capital of Maine, that *Marshall* was the first Chief Justice. Lacking descriptions, the names we learned would be limited to labels for individuals present and salient to teacher and pupil. Under the circumstances, communication would be precarious at best.

The linguistic role of definite descriptions mirrors that of names. A definite description purports to designate an object. As in the case of names, an object can be designated by more than one description. In general, definite descriptions and names are interchangeable in sentential contexts

Socrates is clever.
The philosopher is clever.

Each sentence can be used to ascribe cleverness to the same individual. Letting $Cx = $ "x is clever," the first sentence can be translated into L_p as

$$Cs$$

What of the second? Following Russell [Bertrand Russell, 1872–1970, whose theory of descriptions is advanced in "On Denoting," *Mind* 14 (1905): 479–93] a description,

the philosopher

used in a particular context, aims to designate exactly one object. Were this not so, were there, for instance, no philosopher in the vicinity, or were there more than one, the description (and any sentence containing it) would be defective. An ordinary existentially quantified sentence in L_p captures only the first of these conditions

$$(\exists x)Px$$

The sentence says that there is *at least one* philosopher, leaving open the possibility that more than one object answers to the predicate "is a philosopher." The definite description, in contrast, appears to exclude this possibility. How can we narrow the focus? Suppose we combine the sentence above—"There is at least one philosopher"—with the sentence "There is *at most one* philosopher"

$$(\exists x)(Px \wedge (\forall y)(Py \supset x = y))$$

In pidgin L_p

There is at least one x such that x is a philosopher and, for all y,
if y is a philosopher, then x and y are one and the same.

That is, there is *at least one* philosopher and *at most one*. If Russell is right, if the definite description "the philosopher" is equivalent to "there is at least one philosopher, and at most one," we can translate "The philosopher is clever" as

$$(\exists x)((Px \wedge (\forall y)(Py \supset x = y)) \wedge Cx)$$

The definite description establishes the role of x as the designation of a unique philosopher. All that remains is an attribution of cleverness to the object of that designation.

Language and Thought

"Language disguises thought. So much so, that from the outward form of the clothing it is impossible to infer the form of the thought beneath it, because the outward form of the clothing is not designed to reveal the form of the body, but for entirely different purposes.

"The tacit conventions on which the understanding of everyday language depends are enormously complicated."

[Ludwig Wittgenstein, *Tractatus Logico–Philosophicus*, trans. D. F. Pears and B. F. McGuinness (London: Routledge and Kegan Paul, 1961), § 4.002.]

Consider a slightly more complicated sentence that incorporates a definite description

The young philosopher admires Socrates.

The sentence features a definite description

The young philosopher....

According to Russell, the description implies that there is exactly one young philosopher. Suppose we let $Px = $ "x is a philosopher," $Yx = $ "x is young," and $A①② = $ "① admires ②." To say that there is exactly one young philosopher is to say that there is *at least one*

$$(\exists x)(Px \wedge Yx)...$$

and *at most one*

$$\ldots(\forall y)((Py \wedge Yy) \supset x = y)\ldots$$

Conjoining these yields *exactly one*

$$(\exists x)((Px \wedge Yx) \wedge (\forall y)((Py \wedge Yy) \supset x = y))$$

To this we add that this young philosopher, *x*, admires Socrates

$$\ldots Axs$$

Conjoining this to the definite description, we obtain

$$(\exists x)(((Px \wedge Yx) \wedge (\forall y)((Py \wedge Yy) \supset x = y)) \wedge Axs)$$

Note the pattern of parentheses. The sentence has the form of a conjunction, one of whose conjuncts is a conjunction

$$((p \wedge q) \wedge r)$$

The *p* conjunct is itself a conjunction, and the *q* conjunct is a conditional with a conjunctive antecedent. These relations are marked by means of nested parentheses.

Russell's account of definite descriptions is plausible only so long as we suppose that descriptions are used in contexts that narrow the range of objects over which variables range (see § 4.12). We can speak sensibly of "the young philosopher" because the context makes clear that we intend something like "the young philosopher in that group of philosophers over there." In this context, "the young philosopher" excludes philosophers outside the group. Thus, on Russell's view, "the young philosopher" implies that a range of objects includes exactly one young philosopher; it does not imply that the whole universe of objects includes exactly one young philosopher.

Let us summarize. Definite descriptions of the form "The *F*...," are analyzable into conjunctions: "There is at least one *F*" and "...at most one *F*." These can be translated into $\mathcal{L}_p$ as follows:

there is at least one *F*: $(\exists x)Fx\ldots$
...and at most one *F*: $\ldots \wedge (\forall y)(Fy \supset x = y)$
there is exactly one *F*: $(\exists x)(Fx \wedge (\forall y)(Fy \supset x = y))$

Note that the middle expression above is not a sentence of $\mathcal{L}_p$. It contains, in addition to a dangling connective, a free variable, an occurrence of x that is picked up by a quantifier and thus acquires significance only when the expression is included in the sentence at the bottom of the list.

Identity and Indiscernibility

The philosopher Leibniz (*Discourse on Metaphysics*, § 9) is usually credited with formulating a principle of identity:

(1) *Identity of indiscernibles*: if every predicate true of x is true of y and *vice versa*, then $x = y$.

According to this principle, if x and y are exactly alike in every respect (if x and y are indiscernible), then x and y are one and the same object.

A distinct, but closely related principle, has been advanced by Quine [*From a Logical Point of View* (New York: Harper and Row, 1963), p. 139]:

(2) *Indiscernibility of identicals*: if $x = y$, then every predicate true of x is true of y and *vice versa*.

The principles are not equivalent. Can you see why? How might we go about combining them into one grand principle?

How might we translate into $\mathcal{L}_p$ a self-standing sentence of the form "There is at most one F"? Consider the sentence

There is at most one god.

Letting $Gx = $ "x is a god," the sentence can be translated into $\mathcal{L}_p$ as

$$(\forall x)(Gx \supset (\forall y)(Gy \supset x = y))$$

In pidgin $\mathcal{L}_p$

> For all x, if x is a god, then for all y,
> if y is a god, x and y are identical.

The sentence does not imply that there *is* a god—$(\exists x)Gx$—nor should it. To say that there is at most one is to say that *if* there is an object answering to the predicate "is a god," then there is only one such object.

Identity and Relations

The identity relation is *transitive* and *symmetric*. It is transitive because

> if x = y and y = z, then x = z.

In this respect identity resembles the *greater than* relation

> if x > y and y > z, then x > z.

But the greater than relation is not symmetric. Although

> if x = y, then y = x

it is not true that

> if x > y, then y > x

Not all relations are transitive. Some, like the relation *is next to* are *nontransitive*

> if x is next to y and y is next z, x may or may not be next to z.

Other relations, like the relation *is the mother of*, are *intransitive*

> if x is the mother of y and y is the mother of z, x is not the mother of z.

You might be tempted to translate the sentence differently

$$(\forall x)(Gx \supset \neg(\exists y)(Gy \wedge x \neq y))$$

A sentence expressible in pidgin $\mathcal{L}_p$ as

> For all x, if x is a god, then it's not the case that
> there is a y such that y is a god and x and y aren't identical.

Is this wrong? No. The two $\mathcal{L}_p$ sentences are logically equivalent. (A proof that this is so must await a discussion of transformation rules applying to sentences of $\mathcal{L}_p$ in the next chapter.)

It is now possible to extend these applications of identity to sentences involving numbers greater than one. Consider the sentence

> There are exactly two gods.

The sentence might be analyzed into

> There are at least two gods,
> and at most two gods.

These can be translated into $\mathcal{L}_p$, respectively, as

> there are at least two gods: $(\exists x)(\exists y)((Gx \wedge Gy) \wedge x \neq y)\ldots$

(Note the inclusion of $x \neq y$.)

> ...and at most two gods: $\ldots(\forall z)(Gz \supset (x = z \vee y = z))$

(Note the $\vee$ in $x = z \vee y = z$.) In pidgin $\mathcal{L}_p$

> there are at least two gods: There is an x and a y, such that x and y are gods, and x and y are not one and the same...
>
> ...and at most two gods: ...for all z, if z is a god, z is identical with x or with y.

Conjoining these, we obtain

$$(\exists x)(\exists y)(((Gx \wedge Gy) \wedge x \neq y) \wedge (\forall z)(Gz \supset (x = z \vee y = z)))$$

Two points noted above bear mention here. First, when more than one individual is introduced, we must add, in addition to a distinct quantifier for each, a clause indicating that the individuals are not one and the same individual. The L_p sentence

$$(\exists x)(\exists y)(Gx \wedge Gy)$$

says twice over "there is a god." My saying twice over "there is a desk in my office; there is a desk in my office," is not the same as my saying "there are *two* desks in my office." To do that, I must say: "there is a desk in my office, and there is *another* desk [i.e., a desk not identical with the first] in my office."

Second, in the "...at most two gods" clause, we must say that if z is a god, then z is identical with *either* x *or* y. The sentence has already established that x and y are not identical, so, assuming that identity is transitive, z could not be identical with both.

How might we express in L_p the self-standing sentence

There are at most two gods.

Bearing in mind our earlier rendition of "There is at most one god," and bearing in mind that this sentence does not imply that any god exists, we can translate it as follows:

$$(\forall x)(\forall y)(((Gx \wedge Gy) \wedge x \neq y) \supset (\forall z)(Gz \supset (x = z \vee y = z)))$$

One way to describe the difference between this sentence and the earlier translation of "There are exactly two gods," is that in this sentence (1) universal quantifiers replace the earlier existentials, and (2) a $\supset$ replaces the $\wedge$ as the sentence's "major connective." The earlier sentence is a complex conjunction, and the sentence we have just created is a complex conditional.

Exercises 4.09

Translate the English sentences below into L_p. Let Sx = "x is a sleuth"; Wx = "x is wily"; Px = "x is a person"; $A①②$ = "① admires ②." Use appropriate lowercase letters for names.

1. The sleuth is wily.

2. There is at least one wily sleuth.

3. At most there is one wily sleuth.

4. The wily sleuth admires someone.

5. Everyone admires the wily sleuth.

6. Gertrude admires the wily sleuth.

7. There are exactly two wily sleuths.

8. There are no more than two wily sleuths.

9. If Fenton and Gertrude are sleuths, then there are at least two sleuths.

10. Fenton admires the two sleuths.

11. No one admires every wily sleuth.

12. Someone admires the wily sleuth.

13. At least two people admire Fenton.

14. At most two people admire the sleuth.

15. If Fenton is admired by the wily sleuth, then Fenton is admired by at least one person.

4.11 COMPARATIVES, SUPERLATIVES, AND EXCEPTIVES

By now the importance of the identity concept should be evident. Were we to lack the concept, we would find it impossible to say or think much of what we now say and think. We can extend these observations by investigating the logic of comparatives, superlatives, and exceptives.

Consider a simple comparative in English

Euterpe is wiser than Clio.

Letting $W①② =$ "① is wiser than ②," the sentence is translated into $\mathcal{L}_p$ as

Wec

Consider, however, the sentence

Euterpe is the wisest Muse.

Note, first, that this sentence is not equivalent to the sentence

Euterpe is wiser than any Muse.

As the sentence implies, Euterpe is a Muse. Euterpe cannot be wiser than *any* muse because she cannot be wiser than herself. If Euterpe is the wisest Muse then

Euterpe is wiser than any other Muse
(i.e., any Muse other than Euterpe).

Thus put, we can see our way to translating it into $\mathcal{L}_p$. Using Mx to mean "x is a Muse"

$$Me \wedge (\forall x)((Mx \wedge x \neq e) \supset Wex)$$

In pidgin $\mathcal{L}_p$

Euterpe is a Muse, and for all x, if x is a Muse
other than Euterpe, then Euterpe is wiser than x.

The sentence in both English and in $\mathcal{L}_p$ does not imply that there are any Muses other than Euterpe. It would be true even if Euterpe were the only

Muse. (Indeed, were Euterpe the only muse, the sentence would be trivially true.)

When we say that some object possesses more of a given trait than some other object or collection of objects, we employ a *comparative* construction. When we say that an object possesses more of that trait than any other object, we use a *superlative*. A superlative is a species of comparative. We compare an object belonging to some class to every other member of that class. The wisest Muse is the Muse wiser than any other Muse; the highest peak is the peak higher than any other peak. The wisest Muse need not be the wisest object, and the highest peak need not be the highest object.

Exceptives compare an object belonging to some class of objects to every other member of that class *other than* one or more specified members. Here is a example:

> Clio is the wisest Muse except for Euterpe.

According to this sentence, Clio is the wisest member of the class consisting of every Muse *minus* Euterpe. The sentence does not imply that Euterpe is wiser than Clio, only that Clio is no wiser than Euterpe. The possibility that Clio and Euterpe are tied with respect to wisdom is left open.

Translating the sentence into L_p is straightforward so long as we bear these observations in mind. As in the case of the superlative sentence above, the sentence tells us, first, that both Clio and Euterpe are Muses

$$Mc \land Me$$

We are not entitled to add that Clio and Euterpe are not one and the same individual: $c \neq e$. The sentence may suggest that this is so, but it does not say so explicitly. Instead, we designate a class consisting of all Muses *except* Clio and Euterpe, and indicate that Clio is wiser than any member of that class

$$(\forall x)((Mx \land (x \neq c \land x \neq e)) \supset Wcx)$$

The class of Muses must exclude Clio as well as Euterpe; otherwise the sentence would imply that Clio is wiser than herself. This obliges us to use a conjunction, $x \neq c \land x \neq e$. Assembling both components

$$(Mc \land Me) \land (\forall x)((Mx \land (x \neq c \land x \neq e)) \supset Wcx)$$

In pidgin L_p

Clio is a Muse and Euterpe is a Muse, and for all x, if x is a muse other than Clio and other than Euterpe, then Clio is wiser than x.

Two more sorts of construction bear mention in this context. An example of the first occurs in the English sentence

Only Euterpe is wise.

The sentence expresses the thought that the class of wise things includes just one member, Euterpe. If anything is a member of this class, it must be Euterpe. In $\mathcal{L}_p$, and assuming that $Wx =$ "x is wise"

$$We \wedge (\forall x)(Wx \supset x = e)$$

The same translational pattern applies to sentences featuring complex terms

Euterpe is the only wise Muse.

In $\mathcal{L}_p$

$$(Me \wedge We) \wedge (\forall x)((Mx \wedge Wx) \supset x = e)$$

The translation makes clear the sentence's truth conditions: Euterpe is a wise Muse, and anything that is a wise Muse is Euterpe.

Exercises 4.10

Translate the English sentences below into L_p. Let Sx = "x is a sleuth"; Bx = "x is brave"; $T①②$ = "① is thinner than ②." Use appropriate lowercase letters for names.

1. Fenton is thinner than Chet.

2. Fenton is not thinner than some sleuth.

3. Callie is the thinnest sleuth.

4. No sleuth is thinner than Callie.

5. The thinnest sleuth is Callie.

6. Except for Callie, Fenton is the thinnest sleuth.

7. The thinnest sleuth is brave.

8. The thinnest sleuth is thinner than Chet.

9. If Callie is the thinnest sleuth, then she is brave.

10. If Fenton is a sleuth, he's not the thinnest sleuth.

11. Only Callie is a brave sleuth.

12. Only Callie and Fenton are brave sleuths.

13. The brave sleuth is thinner than Fenton or Chet.

14. If any sleuth is brave, then the thinnest sleuth is brave.

15. The thin sleuth is the brave sleuth

4.12 THE DOMAIN OF DISCOURSE

The variables we employ in L_p range indiscriminately over *all* objects. Consider the English sentence

Everything is lost.

Suppose we translate this sentence into L_p (letting $Lx =$ "x is lost"):

$$(\forall x)Lx$$

This sentence means that everything—literally *everything*: people, stars, prime numbers, grains of sand on the surface of Mars—is lost. Let us call the set of individuals over which the variables in a sentence (or a collection of sentences, or a language) range, the *domain of discourse* for that sentence (collection of sentences, language). The domain of discourse for sentences in L_p, unless otherwise specified, is unrestricted; it includes every object.

In using a natural language like English, we rely heavily on contextual information in deciding how to express our thoughts and how to interpret the utterances of others. We noted the effects of context in our discussion of definite descriptions. If "the philosopher is wise" is correctly analyzed along the lines suggested by Russell ("There is *exactly one* philosopher, x, and x is wise"), then most uses of definite descriptions depend on an implicit narrowing of the domain by speakers and hearers. We might attempt to build these contextual restrictions into the sentence: "the philosopher in the kitchen on 20 January 1994." It is far from clear, however, that contextual information could be made entirely explicit. You will have noticed that "the kitchen" is itself a definite description.

It could turn out that the implicit narrowing we make use of in interpreting language cannot be made fully explicit. Were that so, an explanation of the semantics of natural languages like English would require an appeal to something other than the semantics of individual sentences. Our understanding of the sentence "the philosopher is wise" on a particular occasion could not be explained by assuming that we use our knowledge of the context to "add to" the sentence behind the scenes: no matter how much we beef up the sentence, a grasp of its meaning still requires an understanding of the context. Context makes it plain that when I say "everything is lost," I am not making a claim about absolutely everything; and when you say that everyone is invited to a party at your house, you do not mean, nor does your audience understand you to mean, literally *everyone*.

Even supposing that the role of context is ineliminable, we can and do take steps explicitly to narrow the class of objects over which we take our discourse to range. One narrowing technique is mirrored in L_p. We narrow a claim by specifying that it concerns only a particular class of individuals. You invite *everyone in the office* to a party, for instance. Similarly, we can confine our claim about what is lost to the class of good things

Everything good is lost.

Letting $Gx =$ "x is good"

$$(\forall x)(Gx \supset Lx)$$

If our interest is in producing in L_p sentences whose explicit truth conditions approximate the truth conditions of ordinary English utterances, this device moves us in the right direction, although, inevitably, it falls short.

If the aim of logic is to eliminate all uncertainty, all indeterminacy in language, the aim is realizable only by increasing the distance between logic and ordinary speech. A more modest approach is preferable. We allow the same latitude in the interpretation of sentences in L_p that we allow in the interpretation of English sentences. As a result, the significance of a sentence in L_p will vary from occasion to occasion; but that is so in any case, and it is so for utterances of English sentences. Just as we recognize that "everything" in "everything is lost" can denote different everythings on different occasions, so we recognize that the L_p sentence

$$(\forall x)Lx$$

can range over different objects at different times. In allowing a role for context, we are not lowering our sights. We are merely conceding the inevitable.

The moral: in using L_p, what we take variables to range over depends both on the predicates contained in a sentence and on contextual factors, stated and unstated.

Why not make a virtue of necessity? Suppose we set out to formulate axioms of arithmetic and to prove theorems from those axioms. Our interest would lie only with numbers, their properties, and relations. We might set out to make this clear in *each sentence*. That is, each sentence would explicitly limit the range of variables it included to numbers, just as in the sentence above, we limited the range of variables to good things. After a while, that might begin to seem pointlessly repetitive. For

There is an even prime number

we would write something like

$$(\exists x)(Nx \wedge (Ex \wedge Px))$$

(supposing Nx = "x is a number," Ex = "x is even," and Px = "x is prime"). For

Every number has a successor

we might write

$$(\forall x)(Nx \supset (\exists y)(Ny \wedge Syx))$$

(letting $S①②$ = "① is the successor of ②").

There is another possibility. We might elect to impose a restriction in the metalanguage on the class of objects over which variables in the object language are to be taken to range. Were we to do that, and thus to limit the domain of discourse to the class of natural numbers, we could avoid having to include in every sentence an indication that it applies only to members of the class of numbers. (The natural numbers are the whole numbers 0, 1, 2, 3, 4, ….) The sentence

There is an even prime number

could be translated into $\mathcal{L}_p$ as

$$(\exists x)(Ex \wedge Px)$$

We are concerned only with members of the class of natural numbers, so when we say that something is both even and prime, we mean that some *number* is both even and prime. In restricting the domain, we restrict in advance the values xs and ys can take. In the present case, we restrict those values to the natural numbers.

By restricting the domain in this way, a comparable economy can be realized for any sentence concerned with numbers.

Every number has a successor

can be translated into $\mathcal{L}_p$, assuming a domain restricted to numbers, as

$$(\forall x)(\exists y)Syx$$

Compare this sentence with the version above written without benefit of a restricted domain.

Introducing a restricted domain of discourse is not something to be taken lightly. Restricting a domain requires stage setting: we must make it clear to our intended audience what we have in mind. Domain restricting is useful when we are faced with the prospect of formulating a large number of complex sentences, all of which share a subject matter. Otherwise, it is best to assume an unrestricted domain and let context pick up the slack.

Exercises 4.11

Translate the English sentences below into L_p, first assuming an unrestricted domain of discourse, then assuming a domain restricted to the class of sleuths. Let Sx = "x is a sleuth"; Bx = "x is brave"; $A①②$ = "① admires ②." Use appropriate lowercase letters for names.

1. Every sleuth admires some sleuth.

2. Every sleuth admires some brave sleuth.

3. Some sleuth admires every sleuth.

4. Some brave sleuth admires every sleuth.

5. There is exactly one brave sleuth.

6. There is at most one brave sleuth.

7. The brave sleuth admires Fenton.

8. There are at most two sleuths.

9. Every sleuth admires the brave sleuth.

10. No sleuth admires any sleuth who isn't brave.

4.13 THE SYNTAX OF $\mathcal{L}_p$

A description of the syntax of $\mathcal{L}_p$ parallels the description of the syntax of $\mathcal{L}_s$ in § 2.20. The quantificational structure of $\mathcal{L}_p$ adds a measure of complexity to the task, but its overall character is familiar. We begin by specifying classes of symbols from which sentences of $\mathcal{L}_p$ are be constructed. The class of logical connectives, C, and the class of parentheses, P, are the same in $\mathcal{L}_p$ and $\mathcal{L}_s$

C: truth functional connectives, $\{\neg, \wedge, \vee, \supset, \equiv, \neg\}$;
P: left and right parentheses, $\{(,)\}$.

$\mathcal{L}_s$ includes a collection of sentential constants in addition. These are replaced in $\mathcal{L}_p$ with individual terms, predicate letters, and quantifiers designated by Greek letters: α (*alpha*) designates the set of individual variables; β (*beta*) designates the set of individual constants; Φ (uppercase *phi*) designates the set of predicate letters; and Γ (*gamma*) designates the set of quantifiers.

α: $\{u, v, w, x, y, z\}$
β: $\{a, b, c, \ldots, t\}$
Φ: $\{A, B, C, \ldots, Z\}$
Γ: $\{(\forall u), (\forall v), \ldots, (\forall z), (\exists u), (\exists v), \ldots, (\exists z)\}$

Next, we specify a set of unquantified atomic sentences, Π (*pi*). Π comprises finite strings of symbols consisting of some member of Φ followed by one or more members of β (that is, a predicate letter followed by one or more individual constants)

Π: $\{\phi_0\beta_0, \phi_0\beta_1, \phi_0\beta_2, \ldots, \phi_1\beta_0, \ldots, \phi_0\beta_0\beta_1, \ldots, \phi_n\beta_0\ldots\beta_n\}$

In Π, ϕ_0, $\phi_1, \ldots, \phi_n$ are members of Φ, and β_0, $\beta_1, \ldots, \beta_n$ are members—not necessarily distinct—of β. Expressions $\Phi\beta_0\ldots\beta_n$ must be finite. Π includes, then, Aa, Aab, Faa, Gfh, and so on. Π does not include expressions containing variables—Gxa, Fx, and the like. Members of Π, are simple, self-standing, unquantified sentences.

What about sentences expressing identity relations: "$a = b$," and the like? It will simplify the presentation to treat identity as an ordinary diadic relation designated by a two-place predicate. Thus, we can take "$a = b$" to be expressed by "Iab." Imagine sentences containing the identity relation rewritten in this way. This fiction does not affect the substance of the rules thus far advanced, but it makes their formulation less cumbersome.

It is now possible to provide a *recursive* specification of the set of sentences of $\mathcal{L}_p$. We define this set, Σ (uppercase *sigma*), first by defining the set of sentences lacking quantifiers, those built solely from members of C (the logical constants), P (parentheses), and Π. We then extend Σ to include sentences incorporating quantifiers and variables. As in our characterization of the syntax of $\mathcal{L}_s$, we take a *bounded string* to be a string of symbols enclosed between left and right parentheses. Lowercase Greek *sigmas* (σ_i, σ_j) designate arbitrary members of Σ.

1. Every member of Π is a member of Σ.
2. If σ_i is a member of Σ, then $\neg\sigma_i$ is a member of Σ.
3. If σ_i and σ_j are members of Σ, then $(\sigma_i \wedge \sigma_j)$ is a member of Σ.
4. If σ_i and σ_j are members of Σ, then $(\sigma_i \vee \sigma_j)$ is a member of Σ.
5. If σ_i and σ_j are members of Σ, then $(\sigma_i \supset \sigma_j)$ is a member of Σ.
6. If σ_i and σ_j are members of Σ, then $(\sigma_i \equiv \sigma_j)$ is a member of Σ.

As noted, these rules, like those used to characterize the class of sentences of $\mathcal{L}_s$, are recursive. They begin with a base clause (rule 1), stipulating that members of Π are members of Σ. The remaining rules provide recipes for generating the remaining members of Σ from this stipulated basis. The recursive character of these generative rules is expressed in their taking as inputs members of Σ and producing as outputs new members of Σ.

Thus far, Σ includes $\mathcal{L}_p$ sentences of the form

$$Fa$$
$$(Fa \supset Ga)$$
$$((Gab \wedge Fmn) \vee Ps)$$

As with $\mathcal{L}_s$, we conventionally leave off the outermost parentheses when they enclose a self-standing sentence.

Rules 1–6 produce only sentences of $\mathcal{L}_p$ containing neither quantifiers nor variables. We must now provide a way of generating the remaining sentences. This is accomplished by means of a pair of additional rules:

7. Suppose that $\Phi\beta_0...\beta_n$ is a member of Σ containing β_i, a member of β, and that α_i is a variable not in $\Phi\beta_0...\beta_n$. Let $\Phi\alpha_i$ be the string obtained from $\Phi\beta_0...\beta_n$ by replacing at least one occurrence of β_i by α_i. Then $(\forall\alpha_i)(\Phi\alpha_i)$ is member of Σ.

8. Suppose that $\Phi\beta_0...\beta_n$ is a member of Σ containing β_i, a member of β, and that α_i is a variable not in $\Phi\beta_0...\beta_n$. Let $\Phi\alpha_i$ be the string obtained from $\Phi\beta_0...\beta_n$ by replacing at least one occurrence of β_i by α_i. Then $(\exists\alpha_i)(\Phi\alpha_i)$ is member of Σ.

Rules 1–8 set out conditions whose satisfaction results in sentences of L_p. A final rule makes it explicit that *only* objects satisfying these conditions count as sentences of L_p

9. All and only members of Σ are sentences of L_p.

Although the overall form of the definition is familiar from the recursive characterization of *sentence of L_s* in § 2.20, rules 7 and 8 call for comment. These rules are best understood through examples. Consider the L_p sentence

$$(\forall x)(Fx \supset Gx)$$

Neither Fx nor Gx counts as a member of Σ—hence neither counts as a sentence of L_p: each contains a free variable, x. For the same reason, the expression

$$(Fx \supset Gx)$$

does not count as a sentence either. The expression

$$(Fa \supset Ga)$$

is a member of Σ, and hence a sentence: Both Fa and Ga are members of Π, and so, by rule 1, count as members of Σ. Rule 5 tells us that a $\supset$ flanked by members of Σ is a member of Σ, so the expression above is a member of Σ. Rule 7 envisages the replacement of one or more occurrences of the individual constant a in this expression with a variable. Suppose we replace both occurrences of a with x, thereby obtaining

$$(Fx \supset Gx)$$

To this we add a universal quantifier, $(\forall x)$, to produce a quantified sentence of L_p

$$(\forall x)(Fx \supset Gx)$$

Rule 7 permits the replacement of *one or more* occurrences of *a* with a variable, so the very same input expression could yield

$$(\forall x)(Fx \supset Ga)$$
$$(\forall x)(Fa \supset Gx)$$
$$(\forall y)(Fy \supset Gy)$$
$$(\forall y)(Fa \supset Gy)$$

Rule 7 includes a clause specifying that the variable α_i does not already occur in $\Phi\beta_o\ldots\beta_n$. The prohibition is designed to block strings of the form

$$(\forall x)(\exists x)Fxx$$

By rule 1, the expression *Faa* is a member of Σ, and hence a sentence. An application of rule 8 yields a new member of Σ, the quantified sentence

$$(\exists x)Fxa$$

Suppose, now, we apply rule 7 to this expression. In so doing, the prohibition obliges us replace the remaining occurrence of *a* by some variable *not already present in the sentence*. As a result, rule 7 permits

$$(\forall y)(\forall x)Fxy$$

but not

$$(\forall x)(\exists x)Fxx$$

The same prohibition blocks expressions of the form

$$((\forall x)(Fx \supset (\forall x)Gx))$$

while allowing

$$((\forall x)Fx \supset (\forall x)Gx)$$

and

$$((\forall x)(Fx \supset (\exists y)Gy))$$

Rules 7 and 8, in light of this clause, cannot be used to generate expressions that include quantifiers whose variables match and whose scope overlaps.

Rules 7 and 8 exclude, as well, expressions of the form

$$(\forall x)(Fa \supset Ga)$$

Such expressions incorporate *vacuous* quantifiers, which pick up no variables.

FInally, rule 9 excludes from the set of sentences of L_p expressions not constructed in accord with rules 1–10. Taken together, the nine rules and their accompanying definitions provide necessary and sufficient conditions for sentencehood in L_p. The conditions are *sufficient* in the sense that anything satisfying them counts as a sentence of L_p; the conditions are *necessary* in the sense that *only* items satisfying them count as sentences of L_p.

4.14 THE SEMANTICS OF L_p

The semantics of L_p is an extension of the semantics of L_s. We begin with the notion of an *interpretation*; then, exploiting the recursive account of sentences in L_p, we characterize the notion of a sentence being true under an interpretation.

An interpretation, I, consists of (1) a nonempty, finite or infinite, domain of objects, D; (2) an assignment to each individual constant a member of D; and (3) an assignment to each monadic predicate a set of objects, and to each n-adic predicate a set of ordered n-tuples of objects. Individual constants and predicates can be said to designate the objects and sets of objects assigned to them. Clause (2) allows for the assignment of an object to more than one individual constant, but no individual constant can be ambiguous: no constant can be assigned more than one object. The same object can be assigned to a and to b, but distinct objects cannot be assigned to a. Clause (3) places an analogous restriction on predicates.

A domain might consist of students in a logic class, a collection of stars, or the natural numbers. A domain could just as easily consist of a wildly gerrymandered collection of objects: Socrates, the number two, and this book. The only restriction on sets of objects that make up domains is that the sets must contain at least one object. We leave open what sorts of entity count as objects.

Given an interpretation, I, we can characterize true under I, first for unquantified sentences of L_p, then for sentences containing quantifiers.

1. If σ_i is a member of Π, then σ_i is true under I if and only if the objects I assigns to the individual constants of σ_i are members of the set of ordered n-tuples assigned to the predicate of σ_i.

Recall that σ_i is any member of Π, the set of unquantified simple sentences of $\mathcal{L}_p$, those lacking both variables and connectives. On rule 1, σ_i is true under an interpretation, I, just in case the object or ordered n-tuple of objects assigned by I to the individual constant or constants of σ_i are members of the set of objects or ordered n-tuples of objects assigned to its predicate. (Think of an object by itself as a *one*-tuple.) A sentence like

Fab

is true under I, just in case the objects assigned to "a" and "b" are members of the set of ordered pairs assigned to F, such that a is the first member, b the second.

2. $\neg\sigma_i$ is true under I if and only if σ_i is not true under I.
3. $(\sigma_i \wedge \sigma_j)$ is true under I if and only if σ_i is true under I and σ_j is true under I.
4. $(\sigma_i \vee \sigma_j)$ is true under I if and only if σ_i is true under I or σ_j is true under I, or both.
5. $(\sigma_i \supset \sigma_j)$ is true under I if and only if σ_j is not true under I and σ_i is true under I, or both.
6. $(\sigma_i \equiv \sigma_j)$ is true under I if and only if σ_i and σ_j are both true under I or neither is true under I.

These rules are images of those used in § 2.20 to characterize the notion of truth under an interpretation for $\mathcal{L}_s$. This is unsurprising. This segment of $\mathcal{L}_p$ consists of unquantified atomic sentences and truth functions of atomic sentences.

A reminder: to simplify the presentation, we are treating sentences of the form $a = b$ as predications of the form Iab. We can suppose that the interpretation of I①② remains fixed: on every interpretation, Iab is true just in case a and b codesignate, that is, the object assigned to a is one and the same as the object assigned to b.

Let us see how rules 1–6 apply in a particular case. Suppose we provide an interpretation of $\mathcal{L}_p$ having as its domain the planets. We let constants designate planets and predicates designate sets of n-tuples as follows:

   R: the set of ringed planets
   A: the set of planets having an atmosphere
   L: the set of ordered pairs of planets in which the first planet is
         larger with respect to its mass than the second
   a: Mars
   b: Earth
   c: Uranus

Call this interpretation I^*. Now consider the sentence

$$Lbc$$

According to rule 1, this sentence is true under I^* just in case the objects assigned by I^* to b and c are such that $\langle b,c \rangle$ is a member of the set of ordered pairs assigned by I^* to L. In English: the sentence is true under I^* just in case Earth is larger with respect to its mass than Uranus. (The Earth's mass does not exceed Uranus's, so, under I^*, Lbc is false.)

Consider a more complex sentence

$$(Aa \wedge Rc) \supset \neg Lac$$

By rule 1, we know that Aa is true under I^* if and only if Mars has an atmosphere, and Rc is true if and only if Uranus is ringed. Rule 3 tells us that $(Aa \wedge Rc)$ is true under I^* just in case both Aa and Rc are true. (Both *are* true under the interpretation, so the conjunction is true.) By rule 1, we know that Lac is true if and only if Mars is larger with respect to its mass than Uranus. According to rule 2, $\neg Lac$, is true under I^* just in case Lac is not true. The mass of Uranus is greater than that of Mars, so Lac is not true under the interpretation, and $\neg Lac$ is true. Finally, rule 5 tells us that a conditional sentence is true under an interpretation if and only if either its consequent is true under that interpretation, or its antecedent is false, or both. In the present case, the consequent, $\neg Lac$, is true under I^*, so the sentence as a whole is true under that interpretation.

We now turn to sentences containing quantifiers. Following Mates [Benson Mates, *Elementary Logic*, 2d ed (New York: Oxford University Press, 1972), chap. 4], let us introduce the notion of β-variant interpretations. Suppose that I_i and I_j are interpretations of L_p, and β is an individual constant (i.e., one of the lowercase letters a, b, c,..., t). Then I_i is a β-variant of I_j just in case I_i and I_j differ at most with respect to the object they assign to β. Note: if I_i and I_j differ at most with respect to the object they assign to β, then I_i and I_j have the same domain. Further, on this characterization I_i is a β-variant of itself.

Let us say that $(\forall\alpha)\phi\alpha$ and $(\exists\alpha)\phi\alpha$ are sentences of L_p consisting of a quantifier followed by an expression containing no free variable other than α. Notice that were some variable other than α free in either, neither $(\forall\alpha)\phi\alpha$ nor $(\exists\alpha)\phi\alpha$ would be sentences of L_p; they would contain a free variable not picked up by $(\forall\alpha)$ or $(\exists\alpha)$. One more notational gimmick: let $\phi\alpha/\beta$ be the result of replacing every free occurrence of α in $\phi\alpha$ with an individual constant, β; that is, it is the result of replacing every instance of the variable α freed when we drop $(\forall\alpha)$ or $(\exists\alpha)$ with a constant β.

Suppose $(\forall\alpha)\phi\alpha$ is the sentence

$$(\forall x)(Ax \supset (\exists y)Ry)$$

Here, α is x, and $\phi\alpha$ is the expression minus the universal quantifier, $(\forall x)$:

$$Ax \supset (\exists y)Ry$$

The expression is not a sentence because it contains a free variable, x. Now, $\phi\alpha/\beta$ is the result of replacing this x with some individual constant, a

$$Aa \supset (\exists y)Ry$$

This is a sentence of L_p, and so, given an interpretation, has a truth value.

When a variable freed in this way is replaced with an individual constant, it is replaced by the first individual constant not already present in $\phi\alpha$. For this purpose, we take the class of individual constants to be ordered, a being the first constant, b being the second, c being the third, and so on.

We can now complete the characterization of "true under I" for L_p

7. $(\forall\alpha)\phi\alpha$ is true under I if and only if $\phi\alpha/\beta$ is true under every β-variant of I.
8. $(\exists\alpha)\phi\alpha$ is true under I if and only if $\phi\alpha/\beta$ is true under at least one β-variant of I.
9. σ_i is false under I if and only if σ_i is not true under I.

How are rules 7 and 8 to be understood? First, consider a simple case, given the interpretation, I^*, set out already

$$(\forall x)Rx$$

According to rule 7, this sentence is true under I^* just in case Ra is true under every a-variant of I^*. An a-variant of I^* is an interpretation differing from I^* at most with respect to the individual assigned to the individual constant a. I^* assigns Mars to a. One a-variant of I^* is the interpretation that assigns Earth to a. Another a-variant assigns Uranus to a. Applying rule 7, $(\forall x)Rx$ is true under I^* if and only if Ra is true under *every* a-variant of I^*. Thus, $(\forall x)Rx$ is true under I^* just in case Mars, Earth, and Uranus are each ringed. (Mars and Earth lack rings, so $(\forall x)Rx$ is not true under I^*.)

Vocabulary Expansion

L_p has been endowed with a finite vocabulary: twenty individual constants, twenty-six predicates, and six variables. We might imagine L_p expanded to L_p^*. In addition to the individual constants a, b, c,..., L_p^* allows constants distinguished by subscripts: a_1, a_2, a_3..., a_n, b_1,...b_n,...t_n. This provides us with an infinite number of individual names. We can imagine the stock of individual variables and predicates similarly supplemented.

What are the advantages of L_p^* over L_p? A natural language cannot run out of words. Although English contains a finite number of terms, it provides mechanisms for the creation of endless new terms. By enlarging the vocabulary of L_p, we dramatically increase its usefulness and expressive power, and we bring it into closer alignment with natural languages.

Now consider the sentence discussed earlier

$$(\forall x)(Ax \supset (\exists y)Ry)$$

This sentence, on rule 7, is true under I^* if and only if the sentence

$$Aa \supset (\exists y)Ry$$

is true under every a-variant of I^*. A conditional sentence is true under an interpretation just in case either its antecedent is false or its consequent is true, or both. The consequent of the conditional under consideration is itself a quantified sentence

$$(\exists y)Ry$$

We know from rule 8 that this sentence is true under I^* just in case Rb is true under *some* b-variant of I^*. One b-variant of I^* assigns Uranus to b. If Uranus has rings, Rb is true under that interpretation, hence $(\exists y)Ry$ is true under I^*. Now, taking the sentence as a whole, we can see that it is true under I^* if and only if either it is not true that every planet has an atmosphere, or some planet has rings, or both.

If you have followed all of this, then you can see how the characterization of true under I applies to more complex sentences. We are now ready to push ahead with a discussion of derivations $\mathcal{L}_p$. This will be the focus of chapter five. But first....

4.15 LOGIC AND ONTOLOGY

A tradition running deep in Western philosophy is encapsulated in the idea that language mirrors the world: linguistic categories reflect categories of being. Words name features of the world; features thus designated serve as the words' meanings. The sentence "Socrates is wise," contains a pair of names, "Socrates" and "is wise." The former names Socrates, the latter, the property of being wise. The meaning of "Socrates" is the man, Socrates, and the meaning of "is wise" is the property of being wise, *wisdom.*

How, on such a picture, can we account for falsehood? A false sentence is not meaningless. But the sentence "Socrates flies" corresponds to nothing in the world, and so would seem to lack meaning. The worry can be avoided by supposing that the world contains entities corresponding to "Socrates" and "flies," but that these entities are not configured in the way the sentence says they are configured: Socrates lacks the property.

What of "Athena is wise"? Is the sentence false? No object answering to "Athena" exists. "Athena," then, lacks a meaning. Sentences containing meaningless terms are meaningless, so the sentence must be meaningless. If that seems excessive, perhaps we could find some other entity to serve as the meaning of "Athena": an idea or a concept.

Thoughts like these have led philosophers at various times to propose fundamental linguistic reforms. If we could clean up language, our talk about the world would avoid nonsense and confusion. We would still utter the occasional falsehood, but falsehood is curable in a way that confusion is not. We reserve names like "Socrates" for existing individuals; merely apparent names such as "Athena" and "Euterpe" can be placed in a distinct grammatical category that marks them as designating, not ordinary objects, but extraordinary objects.

The dream is to construct an ideal world-representing language, one in which the fundamental structure of reality is an open book. Perhaps L_p represents a step along the road toward the fulfillment of this dream. More likely, the dream is untenable. We use language to talk about the world among other things. Our beliefs about the world are expressed in our language. We can discover what we believe by listening to ourselves. Other things equal, we prefer true beliefs, true theories about the world, to false beliefs and theories. Our view of the world, if not the world, is reflected in what we say about it.

The idea that the structure of our language mirrors, or ought to mirror, the structure of the world is transformed into the the idea that our language mirrors us: our beliefs and theories about the world are reflected in how we choose to talk. There is reason to regard L_p as particularly revealing in this respect. An unanticipated advantage of pressing L_p into service in the formulation of our ideas is that, in so doing, we can come to see starkly what those ideas do and do not commit us to. We may, as a result, revise our ideas. Or we may decide that L_p is an inadequate medium for their expression.

"To be is to be the value of a bound variable." So says Quine [see, e.g., *From a Logical Point of View* (New York: Harper and Row, 1963), 1–19]. We can ascertain the character of our "ontological commitments" by noting the sorts of object we countenance as values of variables in a language like L_p. In setting up L_p, we left open the character of objects. This is as it should be. What there is, is a matter for science, not logic, to discover. Do atoms and molecules exist? What of medium-sized items: tables, chairs, planets? A more manageable question from the perspective of logic is whether our conception of the world includes a commitment to the existence of these things. If we find it necessary to "quantify over" variables that take atoms, molecules, and planets as values, then we are so committed.

We occasionally discover that it is possible and, for the sake of overall plausibility, desirable to paraphrase objects away. Is there an average American? We sometimes talk as though there were. The average American, we are told, owns two television sets. Thus: "there is exactly one x such that x is an average American...." If this strikes us as silly, we will find ways of paraphrasing away talk about the average American. The question is, how far can we and ought we go in this direction? Might we, for instance, paraphrase away talk about tables and trees, replacing it with talk about clouds of molecules? In such matters we are guided by pragmatic criteria.

We accept paraphrases that tend to make our overall view simpler and more coherent. Our view of the world is simpler and more coherent if we need not suppose that in addition to individual Americans, the world contains another entity, the average American.

Infinity

In ordinary speech, we treat "infinite" as a synonym for "unimaginably large." This is not what mathematicians have in mind when speaking of infinity. An infinite set is not just an unimaginably large set; an infinite number is not just a very large number. Infinities are different in kind and not just degree from ordinary magnitudes.

One way to characterize an infinite set is as follows:

> C is infinite if and only if the elements of C stand in a one–one correspondence with a proper subset of the elements of C.

(C^* is a proper subset of C just in case every element of C^* is an element of C, but there is an element of C that is not an element of C^*.)

The set of natural numbers—0, 1, 2, 3, 4,...—is infinite. There is a one-one correspondence of its elements with a proper subset of its elements! Thus, there is a one-one correspondence between the natural numbers (0, 1, 2, 3, 4,...) and the set of odd numbers (1, 3, 5, 7, 9, ...). If that seems impossible, you are probably trying to imagine how it could work with a finite set. But it cannot work with a finite set, so the comparison is illegitimate.

The set of natural numbers is said to be denumerably infinite. Any set is denumerably infinite if it can be put into one–one correspondence with the set of natural numbers. Some infinite sets are not denumerable. The set of real numbers, for instance, is not denumerable. (The set of real numbers includes the rationals—those numbers expressible as ratios, or "fractions"—and the irrational numbers—those numbers, like Π or $\sqrt{2}$ inexpressible as ratios of two integers.)

Would we realize a comparable economy and coherence by eliminating reference to tables, trees, mountains? What of values? Minds? Properties? Numbers? Is our view of the world improved or muddied by supposing it to contain such entities? Such questions go beyond the scope of this book. The aim here is just to sketch an application of logic not obvious to casual examination, an application whose significance should not be underestimated. Put to work, $\mathcal{L}_p$ provides a mirror, not of the world, but of our considered opinions about the world. Like any mirror, what it reveals may or not meet with our approval. What matters is not that we like what we find, but that we can see our way to something better.

5

Derivations in $\mathcal{L}_p$

5.00 PRELIMINARIES

The construction of derivations in $\mathcal{L}_p$ closely resembles the construction of derivations in $\mathcal{L}_s$. Derivations have the same form, and the same rules apply. Consider, for instance, the sequence

> If Socrates is wise, then he is happy.
> Socrates is wise.
> Therefore, Socrates is happy.

Letting $Wx =$ "x is wise" and $Hx =$ "x is happy," this sequence can be represented in $\mathcal{L}_p$ as

> 1. + $Ws \supset Hs$
> 2. + Ws
> 3. ? Hs

and the conclusion derived in a single step via MP

> 1. + $Ws \supset Hs$
> 2. + Ws
> 3. ? Hs
> 4. Hs 1, 2, MP

The addition of quantifiers need not call for techniques different from those we deployed in $\mathcal{L}_s$. Consider the English sequence below and its counterpart in $\mathcal{L}_p$

> If something is at rest, then everything is at rest.
> Not everything is at rest.
> Therefore, nothing is at rest.

Letting $Rx =$ "x is at rest," we obtain

1. + $(\exists x) Rx \supset (\forall x) Rx$
2. + $\neg(\forall x) Rx$
3. ? $\neg(\exists x) Rx$
4. $\neg(\exists x) Rx$ 1, 2, MT

The application of transformation rules to sentences in L_p is equally straightforward. Let us consider some examples. Consider the sentence of L_p in the first line of the sequence above

$$(\exists x) Rx \supset (\forall x) Rx$$

We can apply *Cond* to this sentence, converting the $\supset$ to a $\vee$, and changing the valence of the antecedent

$$\neg(\exists x) Rx \vee (\forall x) Rx$$

Applying *DeM* to this sentence yields

$$\neg((\exists x) Rx \wedge \neg(\forall x) Rx)$$

The disjunction has been changed to a conjunction, and the valence of each conjunct and the valence of the whole expression have been reversed.

CP and *IP* also apply in L_p just as they do in L_s—as the derivation below illustrates:

1. + $(\forall x) Fx \supset (\exists x) Hx$
2. ? $(\forall x) Fx \supset ((\forall x) Fx \wedge (\exists x) Hx)$
3. $\lceil (\forall x) Fx$
4. | ? $(\exists x) Gx \wedge (\exists x) Hx$
5. | $(\exists x) Hx$ 1, 3, MP
6. $\lfloor (\forall x) Fx \wedge (\exists x) Hx$ 3, 4, $\wedge I$
7. $(\forall x) Fx \supset ((\forall x) Fx \wedge (\exists x) Hx)$ 3–6, CP

In applying transformation rules to quantified expressions, care must be exercised in the manipulation of negation signs. If we apply *DeM* to the quantified disjunction below

$$(\exists x)(Fx \wedge \neg Gx)$$

the *disjunction* changes its valence, so we insert a negation sign *inside* the existential quantifier

$$(\exists x)\neg(\neg Fx \vee Gx)$$

Compare this with a case in which *DeM* is applied to a sentence containing a quantifier that is itself a part of a disjunction

$$(\forall y)\neg((\exists x)Fx \vee Gy)$$

DeM yields

$$(\forall y)(\neg(\exists x)Fx \wedge \neg Gy)$$

The key to understanding the application of transformation rules to sentences in L_p is understanding when they apply to an expression that happens to be quantified and when they apply to an expression containing a quantifier as a part. Consider the sentence

$$(\forall x)(Fx \supset Gx)$$

When we apply *Cond* to this sentence, we apply the rule to an expression that happens to be quantified

$$(\forall x)(\neg Fx \vee Gx)$$

When we apply *Cond* to an expression containing a quantifier as a part

$$(\forall y)((\exists x)Fx \supset Gy)$$

the contained quantifier is affected

$$(\forall y)(\neg(\exists x)Fx \vee Gy)$$

All this makes perfect sense when you think about it. Like much else in logic, it is easier done than said.

─────────── **Exercises 5.00** ───────────

Construct derivations for the $\mathcal{L}_p$ sequences below.

1. + $(\forall x)(Fx \supset Gx)$
   + $(\forall x)(Gx \lor \neg Fx) \supset (\exists y)Fy$
   ? $(\exists y)Fy$

2. + $(\exists x)Fx \land (\exists x)Gx$
   + $(\exists x)Hx \supset \neg(\exists x)Gx$
   ? $\neg(\exists x)Hx$

3. + $(\exists y)Hy$
   + $((\forall x)Fx \lor (\exists y)Hy) \supset Fa$
   ? Fa

4. + $(\exists x)Fx \supset (\exists x)Gx$
   + $\neg(\exists x)Gx \lor (\forall x)Fx$
   ? $(\exists x)Fx \supset (\forall x)Fx$

5. + $(\forall x)(Fx \land (Gx \lor Hx))$
   ? $(\forall x)((Fx \land Hx) \lor (Fx \land Gx))$

6. + $(\forall x)(Fx \supset Gx)$
   ? $(\forall x)\neg(Fx \land \neg Gx)$

7. + $(\forall x)(Fx \supset Gx) \equiv (\exists x)Fx$
   ? $(\forall x)(Fx \supset Gx) \supset (\exists x)Fx$

8. + $(\exists x)((Fx \land \neg Gx) \lor \neg Gx)$
   ? $(\exists x)((Fx \lor \neg Gx) \land \neg Gx)$

9. + $(\exists x)Fx \supset (\exists x)(Fx \land Gx)$
   + $\neg(\exists x)(Fx \land Gx) \supset (\exists x)Fx$
   ? $(\exists x)(Fx \land Gx)$

10. + $(\forall x)Fx \supset \neg(\exists y)Gy$
    + $\neg(\exists x)Hx \supset (\exists y)Gy$
    ? $(\forall x)Fx \supset (\exists x)Hx$

11. + $\neg(\forall x)Hx$
    + $((\exists x)Fx \lor (\exists x)Gx) \supset (\forall x)Hx$
    ? $\neg(\exists x)Fx$

12. + $(\forall x)Fx \supset (Ga \land Ha)$
    + $((\forall x)Fx \supset Ha) \supset (\exists x)Jx$
    ? $(\exists x)Jx$

13. + $(\forall x)((\exists y)(Fy \land Gy) \lor (Hx \land Jx))$
    ? $(\forall x)(((\exists y)(Fy \land Gy) \lor Hx) \land (Jx \lor (\exists y)(Fy \land Gy)))$

14. + $(\forall x)(Fx \supset Gx))$
    ? $(\forall x)(\neg Gx \supset \neg Fx) \land (\forall x)(Gx \lor \neg Fx)$

15. + $\neg((\forall x)Fx \lor Ga) \lor (\exists y)(Gy \land Hy)$
    ? $(\forall x)Fx \supset Ga$

5.01 QUANTIFIER TRANSFORMATIONS

In all likelihood, you have already noticed that English sentences that require quantifiers when they are translated into $\mathcal{L}_p$ can be translated in more than one way. Consider the sentence

Every philosopher is wise.

It is natural to translate this sentence into L_p as (assuming: Px = "x is a philosopher" and Wx = "x is wise")

$$(\forall x)(Px \supset Wx)$$

But consider: the original English sentence could be paraphrased, albeit awkwardly, as

It's not the case that there is a philosopher who isn't wise.

It is most natural to translate *this* sentence into L_p as

$$\neg(\exists x)(Px \land \neg Wx)$$

How is this sentence related to the original conditional sentence above?

Before answering this question, let us look at some simpler sentences couched in pidgin L_p. (In these sentences, we use αs and ϕs [*alphas* and *phis*] to represent arbitrary terms.) Consider the sentence

All αs are ϕ

and a companion sentence

It's not the case that some αs are not ϕ.

Could these sentences have different truth conditions? If not, they are paraphrases of one another. And if the quantificational structure of L_p corresponds to that of English, the proto-L_p sentence

$$(\forall \alpha)\phi$$

must be logically equivalent to

$$\neg(\exists \alpha)\neg \phi$$

Are these expressions (or rather the pairs of sentences corresponding to them) equivalent? Yes—although a proof of this equivalence will not be offered here.

We can formulate a rule for quantifier transformation that reflects this equivalence

$$(QT) \quad (\forall\alpha)\phi \dashv\vdash \neg(\exists\alpha)\neg\phi$$

The rule tells us that it is permissible to replace a universal quantifier with an existential quantifier flanked by negation signs.

Let us return to the sentences we considered earlier. Recall that we had discovered that although it is natural to translate the English sentence

Every philosopher is wise

as

$$(\forall x)(Px \supset Wx)$$

a case could be made for translating it using an existential quantifier

$$\neg(\exists x)(Px \wedge \neg Wx)$$

Suppose we apply the rule advanced above to the universally quantified original, that is, we replace the universal quantifier with an existential quantifier flanked by negation signs

$$\neg(\exists x)\neg(Px \supset Wx)$$

The result is an odd-looking existentially quantified conditional. We can get from this sentence to our existentially quantified alternative sentence by applying *Cond*

$$\neg(\exists x)\neg(\neg Px \vee Wx)$$

and then *DeM*

$$\neg(\exists x)(Px \wedge \neg Wx)$$

Assuming the quantifier transformation rule formulated above, we can derive one sentence from the other; and assuming that our derivation rules are truth preserving, this amounts to a proof that the sentences are logically equivalent in $\mathcal{L}_p$, just as they are in English.

Rule *QT* was described as permitting the replacement of a universal quantifier by an existential quantifier flanked by negation signs. Although this description is perfectly correct, it is misleading. In chapter three we decided to interpret negation signs appearing in rules as instructions to *reverse the valence* of sentences introduced by applications of the rule. In the present case, this means that when we replace a universal quantifier with an existential quantifier, we must (1) reverse the valence of the quantifier, and (2) reverse the valence of the expression to its immediate right. The rule is a two-way rule, so we can read it as licensing the substitution of a universal quantifier for an existential quantifier provided the same two conditions are satisfied.

Suppose we set out to replace the universal quantifier in

$$\neg(\forall x)(Fx \supset Gx)$$

with an existential quantifier. The rule permits the replacement, provided we (1) reverse the valence of the quantifier (changing it in this case from negative to positive), and (2) reverse the valence of the expression to its immediate right

$$(\exists x)\neg(Fx \supset Gx)$$

Again, we can apply the rule in the opposite direction as well. Take the existentially quantified L_p sentence

$$(\exists x)\neg(Fx \wedge \neg Gx)$$

We are permitted replace the existential quantifier with a universal quantifier if we simultaneously reverse its valence and the valence of the expression to its immediate right

$$\neg(\forall x)(Fx \wedge \neg Gx)$$

Rule *QN*, then, licenses transformations of the following sorts:

$$(\forall \alpha)\phi \quad \dashv\vdash \quad \neg(\exists \alpha)\neg\phi$$
$$(\forall \alpha)\neg\phi \quad \dashv\vdash \quad \neg(\exists \alpha)\phi$$
$$\neg(\forall \alpha)\phi \quad \dashv\vdash \quad (\exists \alpha)\neg\phi$$
$$\neg(\forall \alpha)\neg\phi \quad \dashv\vdash \quad (\exists \alpha)\phi$$

These reflect a comparable pattern in English

All αs are ϕ $\dashv\vdash$	It's not the case that some αs are not ϕ
All αs are not ϕ $\dashv\vdash$	It's not the case that some αs are ϕ
Not all αs are ϕ $\dashv\vdash$	Some αs are not ϕ
Not all αs are not ϕ $\dashv\vdash$	Some αs are ϕ

If these English equivalences hold, then QT insures that quantifier transformation in $\mathcal{L}_p$ corresponds to *some/all* transformations in English.

What of sentences containing more than one quantifier? Consider the sentence

$$(\forall x)(\exists y)((Fx \wedge Gy) \supset Hxy)$$

Imagine that we want to change both quantifiers to their opposites. Although it does not matter which quantifier we transform first, we can transform *only one quantifier at a time.* Suppose we begin by substituting an existential quantifier for the universal quantifier. We change the quantifier, its valence, and the valence of the expression to its immediate right. Here the expression to the immediate right is itself a quantifier

$$\neg(\exists x)\neg(\exists y)((Fx \wedge Gy) \supset Hxy)$$

Now we can apply rule QT to the second existential quantifier, converting it to a universal quantifier. Again, we change the quantifier, its valence, and the valence of the expression to its immediate right

$$\neg(\exists x)(\forall y)\neg((Fx \wedge Gy) \supset Hxy)$$

If we so choose, we can now apply additional transformation rules to the remainder of the expression to yield further logically equivalent sentences.

Let us restate rule QT and then apply it and the rules used in the construction of derivations in $\mathcal{L}_s$ in some exercises

$$(QT) \quad (\forall \alpha)\phi \dashv\vdash \neg(\exists \alpha)\neg\phi$$

Exercises 5.01

Construct derivations for the $\mathcal{L}_p$ sequences below using rule QT.

1. $+ (\exists x)(Fx \land Gx)$
   $? \neg(\forall x)(Fx \supset \neg Gx)$

2. $+ (\forall x)(Fx \supset Gx)$
   $? \neg(\exists x)(Fx \land \neg Gx)$

3. $+ (\forall x)((Fx \land Gx) \supset Hx)$
   $? \neg(\exists x)(Fx \land (Gx \land \neg Hx))$

4. $+ (\exists x)((Fx \land Gx) \lor \neg Hx)$
   $? \neg(\forall x)(Hx \land (Fx \supset \neg Gx))$

5. $+ (\forall x)(Fx \land Gx)$
   $? \neg(\exists x)((Fx \land Gx) \supset (Fx \supset \neg Gx))$

6. $+ (\forall x)((Fx \supset Gx) \land Hx)$
   $? \neg(\exists x)((\neg Hx \lor Fx) \land (\neg Hx \lor \neg Gx))$

7. $+ (\exists x)Fx$
   $? (\forall x)\neg Fx \supset (\exists x)Fx$

8. $+ (\forall x)Fx$
   $? \neg((\forall x)Fx \supset \neg(\forall x)Fx)$

9. $+ (\exists x)Fx \supset (\forall x)\neg Fx$
   $? (\forall x)\neg Fx$

10. $+ (\exists x)\neg Fx$
    $? \neg(\neg(\forall x)Fx \supset (\forall x)Fx)$

11. $+ (\forall x)((Fx \land Gx) \supset Hx)$
    $? \neg(\exists x)(Fx \land (Gx \land \neg Hx))$

12. $+ (\forall x)(Fx \supset Gx)$
    $? \neg(\exists x)(\neg Gx \land Fx)$

13. $+ \neg(\forall x)(Fx \supset Gx)$
    $? (\exists x)(Fx \land \neg Gx)$

14. $+ (\forall x)(Fx \supset (Gx \supset Hx))$
    $? \neg(\exists x)((Fx \land Gx) \land \neg Hx)$

15. $+ \neg(\exists x)((Fx \lor Gx) \land Hx)$
    $? (\forall x)((\neg Fx \land \neg Gx) \lor \neg Hx)$

5.02 UNIVERSAL INSTANTIATION (*UI*)

The derivation rules introduced in chapter three supplemented by rule *QT* are not enough to enable us to construct derivations that establish the validity of every valid sequence expressible in $\mathcal{L}_p$. We know that a sequence, $\langle P, c \rangle$, is valid just in case its conclusion, c, is not false if its premises, P, are true. The classic syllogism below is plainly valid:

> All people are mortal.
> Socrates is a person.
> Therefore, Socrates is mortal.

If we translate the sequence into $\mathcal{L}_p$, we cannot derive the conclusion from the premises using only our derivation rules plus rule *QT*

> 1. $+ \ (\forall x)(Px \supset Mx)$
> 2. $+ \ Ps$
> 3. $? \ Ms$

The sequence might suggest an application of *MP*: the first sentence incorporates a conditional, the second contains a predicate found in the antecedent of the conditional, and the conclusion includes a predicate matching a predicate in the consequent of that conditional. *MP* permits the derivation of the consequent of a conditional given its antecedent. *Ps* is not the antecedent of $Px \supset Mx$; however, nor is *Ms*, its consequent, for that matter. Further, the conditional expression is bound by a quantifier. In applying *MP*, we would be violating the general restriction on rules of inference: rules of inference apply only to *whole sentences*.

There is something right about the thought that a proof for the validity of the sequence above involves an application of *MP*. Consider the first sentence in the sequence, "All people are mortal"

$$(\forall x)(Px \supset Mx)$$

If all people are mortal, then it surely follows that if Socrates is a person, then Socrates is mortal. It follows as well that if Euterpe is a person, Euterpe is mortal; and if Babar is a person, then Babar is mortal; and so on. That is, if the sentence above is true, then the following sentences must be true:

$$Ps \supset Ms$$
$$Pe \supset Me$$
$$Pb \supset Mb$$

Let us call these sentences *instantiations* of the original universally quantified sentence. An instantiation results when we (1) drop a universal quantifier and (2) replace the variables that it bound (in this case the x) with an individual constant (s, e, and b in the sentences above).

We can introduce an instantiation rule, *universal instantiation* (*UI*), to capture this pattern of reasoning

$$(UI) \quad (\forall \alpha)\phi \vdash \phi\alpha/\beta$$

This formulation of *UI* calls for some comment. In the rule, $(\forall \alpha)\phi$ represents any universally quantified expression and ϕ stands for the expression less the quantifier $(\forall \alpha)$. $\phi\alpha/\beta$ represents the expression resulting when we (1) drop the universal quantifier, $(\forall \alpha)$, and (2) replace each instance of the variable it bound, each α, with some individual constant, β. Let us look at some examples.

If we apply *UI* to the sentence

$$(\forall x)(Fx \supset Gx)$$

we drop the universal quantifier, $(\forall x)$, and replace the variables it bound, the xs, with some individual constant. *Which* individual constant we choose will depend on the circumstances. Each of the sentences below results from an application of *UI* to the sentences above:

$$Fs \supset Gs$$
$$Fe \supset Ge$$
$$Fb \supset Gb$$

Let us pause and complete the derivation we were obliged to leave off earlier

1.	+ $(\forall x)(Px \supset Mx)$	
2.	+ Ps	
3.	? Ms	
4.	$Ps \supset Ms$	1, UI
5.	Ms	2, 4, MP

In this case, when we drop the universal quantifier, we change each occurrence of the variable it bound to an s, so the resulting sentence can be used with the sentence in line 2 and *MP* to derive the conclusion. This application of *UI* echoes our own reasoning about the original syllogism: suppose all people are mortal; then, if Socrates is a person, he's mortal (here we instantiate the generalization); Socrates is a person; therefore Socrates is mortal.

Let us look at a few more applications of rule *UI*. Consider the sentence

$$(\forall x)((\exists y)Fy \supset Gx)$$

If we apply *UI* to this sentence, dropping the universal quantifier, $(\forall x)$, the existential quantifier, $(\exists y)$, and the variable it binds, y, remain unaffected

$$(\exists y)Fy \supset Ga$$

When the quantifer is dropped, the variable it bound can be replaced by *any* individual constant. In the example above, the constant a is used, but any other would have done as well. The constant we choose depends on characteristics of the sequence. Thus, if we apply *UI* to the sentence below

$$(\forall y)(Fy \supset Gh)$$

we might want to change the variable bound by the quantifier to an h

$$Fh \supset Gh$$

although that will depend in part on other features of the sequence in which the derivation occurs.

Rule *UI* does not permit the replacement of variables freed when a universal quantifier is dropped with another variable. Were we to do so, derivations could include strings of symbols that were not sentences because they contained free variables. In dropping the $(\forall y)$ quantifier in the sentence above, we must replace the variable it bound with an individual constant, one of the lowercase letters $a, b, c, \ldots, t$.

Rule *UI* requires that when a constant replaces a variable, it does so consistently. In the sentence

$$(\forall z)(((Fz \vee Gz) \wedge Hz) \supset Jz)$$

we can drop $(\forall z)$ and replace every occurrence of z with any constant, provided we replace each z with the *same* constant

$$((Fa \vee Ga) \wedge Ha) \supset Ja$$

Thus, rule *UI* yields the sentence above from the original sentence, but not the sentence below:

$$((Fa \lor Gb) \land Ha) \supset Ja$$

Rule *UI*, like the rules of inference, applies only to whole sentences and not to sentence fragments. If the scope of a universal quantifier is limited to a part of a sentence or to a sentence that is itself part of a larger sentence, rule *UI* does not apply. These restrictions are illustrated in the sentences below:

$$\neg(\forall x)(Fx \supset Gx)$$
$$(\forall x)(Fx \supset Gx) \supset (\exists y)(Fy \land Gy)$$

In the first sentence, the quantifier is negated, so part of the sentence in which it occurs is outside its scope—namely, the ¬. In the second sentence, the universal quantifier includes within its scope only the antecedent of a more inclusive sentence. In neither case does *UI* permit the dropping of the universal quantifier, $(\forall x)$.

Exercises 5.02

Construct derivations for the L_p sequences below, making use of rule *UI* where necessary.

1. $+ (\forall x)(Fx \supset Gx)$
   $+ \neg Ga$
   $? \neg Fa$

2. $+ (\forall x)(Fx \supset Gx)$
   $+ (\forall x)(Gx \supset Hx)$
   $? Fa \supset Ha$

3. $+ (\forall x)(Fx \supset Gx)$
   $+ (\forall x)\neg(Gx \land \neg Hx)$
   $? Ha \lor \neg Fa$

4. $+ (\forall x)(Fx \supset Gx)$
   $+ \neg(\exists x)(Gx \land \neg Hx)$
   $? \neg(Fa \land \neg Ha)$

5. $+ (\forall x)Fx \supset (\forall x)Gx$
   $+ \neg(\exists x)\neg Fx$
   $? Ga$

6. $+ (\forall x)Fx \supset (\forall x)Gx$
   $+ (\forall x)\neg Gx$
   $? (\exists x)\neg Fx$

7. $+ (\forall x)(\neg Fx \supset Hx)$
   $+ (\forall x)(Hx \supset Gx)$
   $? (\exists x)(Fx \lor Gx)$

8. $+ (\forall x)Fx$
   $+ (\forall x)(Fx \supset Gx)$
   $? (\exists x)(Fx \land Gx)$

9. $+ (\forall x)(Fx \supset Gx)$
   $+ (\forall x)(Fx \lor (\exists y)Jy)$
   $+ (\exists y)Jy \supset (\forall x)Hx$
   ? $(\exists x)(Gx \lor Hx)$

10. $+ \neg(\exists x)(\neg Fx \lor Hx)$
    $+ (\forall x)(Jx \supset Gx)$
    $+ (\forall x)(Fx \supset Jx)$
    ? $(\exists x)(Fx \land Gx)$

11. $+ (\forall x)(Fax \supset Gx)$
    $+ \neg(\exists x)(Gx \land Fxc)$
    ? $\neg(\forall x)Fxc$

12. $+ (\forall x)(Fx \supset Gx)$
    $+ \neg(\exists x)(Gx \land \neg Hx)$
    ? $(\forall x)Fx \supset Hb$

13. $+ (\forall x)(Fx \supset Gx)$
    $+ \neg(\exists x)(\neg Gx \land Hx)$
    ? $(\exists x)(Fx \land \neg Gx)$

14. $+ (\forall x)(\neg Fxa \supset Gax)$
    $+ \neg(\exists x)Gxb$
    ? $(\exists x)Fxa$

15. $+ (\forall x)(\forall y)(Fxy \lor Fyx)$
    $+ (\forall x)(\neg(\exists y)(Fyx \land \neg Gx) \land \neg(\exists y)Fxy)$
    ? $(\exists x)Gx$

5.03 EXISTENTIAL GENERALIZATION (*EG*)

Rule *UI* permits us to drop universal quantifiers in derivations in a way that mimics ordinary patterns of inference of the form

> <u>All horses are quadrapeds.</u>
> Therefore, if Bucephalus is a horse,
> Bucephalus is a quadraped.

Representing this sequence in $\mathcal{L}_p$ (and assuming that $Hx =$ "x is a horse" and $Qx =$ "x is a quadraped")

1. $+ (\forall x)(Hx \supset Qx)$
2. ? $Hb \supset Qb$
3. $\quad Hb \supset Qb$ 1, UI

UI is an instantiation rule. That is, it licenses inferences about instances of generalizations from sentences expressing those generalizations. We

sometimes reason in the other direction—from sentences about instances to
sentences expressible in L_p via quantifiers

> Bucephalus is a horse.
> Therefore, something is a horse.

This sequence can be translated into L_p as

>   1. $+ Hb$
>   2. ? $(\exists x)Hx$

We can introduce a rule, *existential generalization* (*EG*), that permits
inferences of this sort

$$(EG) \quad \phi\alpha/\beta \vdash (\exists\alpha)\phi$$

Again, $\phi\alpha/\beta$ represents a sentence containing one or more occurrences of an
individual constant, β. $(\exists\alpha)\phi$ is the sentence that results from (1) the
replacement of one or more occurrences of β by occurrences of a variable α
and (2) the addition of an existential quantifier that captures every
occurrence of α in ϕ.

Rule *EG* permits the derivation of the conclusion of the sequence above

>   1. $+ Hb$
>   2. ? $(\exists x)Hx$
>   3. $(\exists x)Hx$ 1, EG

The rule permits, as well, derivations that reflect the pattern of inference
exhibited in the English sequence below

> All horses are quadrapeds.
> Bucephalus is a horse.
> Therefore, something is a quadraped.

The sequence can be translated into L_p and proven valid using *EG*

>   1. $+ (\forall x)(Hx \supset Qx)$
>   2. $+ Hb$
>   3. ? $(\exists x)Qx$
>   4. $Hb \supset Qb$ 1, UI

5.	Qb	2, 4, MP
6.	$(\exists x)\,Qx$	5, EG

Rule EG, like UI, applies only to whole sentences, not to sentences that are themselves parts of other sentences. When we add an existential quantifier, its scope must include the whole sentence, not a part of the sentence to which it is added. The sequence below illustrates both correct and incorrect applications of rule EG

1.	+ $Fa \wedge Ga$	
2.	+ $\neg(Fa \wedge Ga)$	
3.	$(\exists x)(Fx \wedge Gx)$	1, EG
4.	$(\exists x)\neg(Fx \wedge Gx)$	2, EG
5.	$(\exists x)Fx \wedge Ga$	1, EG (*not permitted*)
6.	$\neg(\exists x)(Fx \wedge Gx)$	2, EG (*not permitted*)

In lines 5 and 6, EG has been applied to sentences that are themselves parts of larger sentences. In neither case does the quantifier include in its scope the whole sentence to which it is added.

EG requires that when an individual constant is converted to a variable, the variable be picked up by the added existential quantifier. Given the sentence

$$(\forall x)(Fx \supset Ga)$$

EG would not permit the addition of an $(\exists x)$ and the conversion of a to x

$$(\exists x)(\forall x)(Fx \supset Gx)$$

In this case the x in Gx is picked up, not by the existential quantifier, but by a universal quantifier, $(\forall x)$, already present in the sentence. EG *would* permit the following sentence to be inferred:

$$(\exists y)(\forall x)(Fx \supset Gy)$$

Here the y in Gy is picked up by the new existential quantifier.

Although EG does not permit an inference from

$$Fa \wedge Gb$$

to the sentence

$$(\exists x)(Fx \wedge Gx)$$

(distinct individual constants, a and b, are converted to a single variable), *EG does* permit an inference from the sentence

$$Fa \wedge Ga$$

to the sentence

$$(\exists x)(Fx \wedge Ga)$$

In this case only one of the constants, a, is converted to a variable and picked up by the new quantifier. Is this an oversight?

Consider the English sequence

<u>Euterpe admires herself.</u>
Therefore, someone admires Euterpe.

The sequence is valid, and its validity can be captured in $\mathcal{L}_p$ provided we interpret *EG* as permitting the addition of an existential quantifier to a sentence, ϕ, containing instances of an individual constant, β, without converting every instance of β in ϕ to a variable, α, picked up by the added quantifier. We can represent this sequence in $\mathcal{L}_p$ (restricting the domain to persons and letting $A①② =$ "① admires ②")

1. $+$ Aee
2. ? $(\exists x)Axe$
3. $(\exists x)Axe$ 1, EG

EG allows us to add an existential quantifier without requiring that we convert every instance of the individual constant, e, to a variable, x.

──────Exercises 5.03──────

Construct derivations for the $\mathcal{L}_p$ sequences below,
making use of rule EG where necessary.

1. $+ (\forall x)(Gx \supset Fx)$
   $+ Ga$
   $+ (\exists x)Fx \supset (\forall x)(Gx \supset Hx)$
   ? $(\exists x)Hx$

2. $+ Fa$
   $+ (\exists x)Fx \supset (\forall x)(Gx \vee Hx)$
   $+ (\exists x)(Gx \vee Hx) \supset Ha$
   ? $(\exists x)Hx$

3. $+ (\forall x)(Fx \supset Gx)$
   $+ (\forall x)(Gx \supset (Hx \wedge Jx))$
   ? $(\forall x)Fx \supset (\exists x)(Gx \wedge Hx)$

4. $+ (\exists x)Fx \supset (\forall x)(Gx \supset Hx)$
   $+ (\forall x)(Fx \supset Gx)$
   ? $Fa \supset (Ga \wedge Ha)$

5. $+ \neg(\exists x)(\neg Fx \vee Hx)$
   $+ (\forall x)((\exists y)(Gy \wedge Hy) \supset Jx)$
   ? $(\forall x)(Fx \supset Gx) \supset (\exists x)(Fx \wedge Jx)$

6. $+ (\forall x)(\neg Gx \supset Hx)$
   $+ (\forall x)\neg(Fx \supset Hx)$
   ? $(\exists x)(Fx \wedge Gx)$

7. $+ (\forall x)(Hx \supset Fx)$
   $+ (\exists x)(Hx \vee Kx) \supset (\forall x)Gx$
   $+ (\forall x)(Hx \wedge Jx)$
   ? $(\exists x)(Fx \wedge Gx)$

8. $+ (\forall x)(Fx \wedge Hx)$
   $+ (\exists x)(Gx \vee Ix) \supset Ja$
   $+ Ja \supset (\forall x)(Gx \supset \neg Fx)$
   ? $\neg(\forall x)(Fx \supset Gx)$

9. $+ (\exists x)(\neg Fx \vee \neg Gx) \supset (\forall x)(Hx \supset Jx)$
   $+ \neg(\exists x)(Hx \wedge \neg Jx) \supset (Fa \wedge Ga)$
   ? $(\exists x)(Fx \wedge Gx) \supset Fa$

10. $+ (\forall x)(Gx \supset \neg Hx)$
    $+ (\forall x)(Fx \vee Gx)$
    ? $(\forall x)Hx \supset (\exists x)Fx$

11. $+ (\forall x)(Fx \supset \neg Gx)$
    $+ \neg(\exists x)\neg Fx$
    ? $(\exists x)(Fx \wedge \neg Gx)$

12. $+ (\forall x)(Fx \supset Gx)$
    $+ \neg Gc$
    ? $(\exists x)\neg Fx$

13. $+ (\forall x)(\forall y)(Fxy \supset Gx)$
    $+ (\forall x)(\forall y)(Gx \supset Hxy)$
    ? $(\forall x)Fxa \supset (\exists x)Hxc$

14. + $(\forall x)(Fx \supset Gx)$
    + $\neg(\exists x)(\neg Gx \wedge Fx) \supset Hac$
    ? $(\exists x)(\exists y)Hxy$

15. + $(\forall x)(\forall y)(Fxy \vee Fyx)$
    + $(\exists x)(\exists y)Fxy \supset (\forall x)Gxb$
    ? $(\exists x)(\exists y)Gxy$

5.04 EXISTENTIAL INSTANTIATION (*EI*)

Rule *EG* permits us to add existential quantifiers to sentences and convert constants to variables bound by those quantifiers. In so doing, it licenses inferences of the form

> Euterpe is admirable.
> Therefore, someone is admirable.

Consider a sequence that moves in the opposite direction

> Someone is admirable.
> Therefore, Euterpe is admirable.

This sequence is clearly invalid. From the fact that *someone* is admirable, we cannot infer that Euterpe, a particular person, is admirable.

What *can* be inferred from a sentence of the form "Something is *F*"? If the sentence is true, we know that *F* must be true of *some* particular thing, even though we cannot say *which* thing it is. We can reason as follows: if something is *F*, then *F* is true of something; call that something "Wayne." In so reasoning we introduce an *arbitrary name*, "Wayne," that serves solely as a designation for the object, whichever object it is, of which *F* is true. (If there is more than one object of which *F* is true, "Wayne" designates an arbitrary object in this collection.) The point will become clearer as we consider applications of the rule for *existential instantiation* (*EI*) set out below:

$$(EI) \quad (\exists \alpha)\phi \vdash \phi \alpha / \beta$$

Restriction: β does not already occur in an active sentence, in the conclusion, or in ϕ. (A sentence is active when it is not preceded by a ? or when it occurs within matched brackets, $\lceil$ and $\lfloor$.)

Rule *EI* is founded on the idea that if we can show that something holds of at least one object, we have shown that there is some particular object of which it holds. The rule permits us to drop an existential quantifier and replace instances of the variable it bound with a constant that functions, in the context of the derivation, as an arbitrary name. The restriction on the rule insures that the constant selected does have this function. Consider the English sequence

> All quarks have charm.
> <u>Something is a quark.</u>
> Therefore, something has charm.

The sequence is valid. Were we to spell this out, we could do so as follows: if something is a quark, then there is at least one object that is a quark; call this object "Becky." If all quarks have charm, then if Becky is a quark, she has charm. If Becky has charm, then, something has charm.

This line of reasoning could be spelled out in $\mathcal{L}_p$ as follows:

1.	+ $(\forall x)(Qx \supset Cx)$	
2.	+ $(\exists x)Qx$	
3.	? $(\exists x)Cx$	
4.	Qb	2, *EI*
5.	$Qb \supset Cb$	1, *UI*
6.	Cb	4, 5, *MP*
7.	$(\exists x)Cx$	6, *EG*

In dropping the existential quantifier in line 4, we satisfy the restriction on *EI* and insure that b functions as an arbitrary name, that is, as the name of some object or other of which it is true that it is a quark. The universally quantified sentence in line 1 holds of every object, so it holds of b.

This derivation illustrates the importance of the *order* in which rules for dropping quantifiers are applied. Had we reversed the order of lines 4 and 5, had we first dropped the universal quantifier, replacing the variables it bound with b, and then dropped the existential quantifier, replacing the variable it bound with b, we would have violated the restriction on *EI*. Why does the order in which quantifiers are dropped matter? Rule *UI* permits us to move from a claim about every object to a claim about some particular object. Rule *EI* permits us to move from a claim about some unspecified object to a claim about that object arbitrarily designated. If an individual constant appears in a premise, in an active sentence that occurs earlier in a derivation, or in the conclusion, its use as an arbitrary designation is preempted. Were we to reverse the order of lines 4 and 5, the constant b would not function as an arbitrary name.

The complicated restriction on rule *EI* is designed to insure that constants introduced when an existential quantifier is dropped serve as arbitrary names. Consider the invalid sequence

> Something is red all over.
> <u>Something is green all over.</u>
> Therefore, something is red and green all over.

We can represent this sequence in L_p (assuming $Rx =$ "x is red all over" and $Gx =$ "x is green all over")

> 1. $+$ $(\exists x)Rx$
> 2. $+$ $(\exists x)Gx$
> 3. ? $(\exists x)(Rx \wedge Gx)$

We might attempt to construct a derivation of this sequence along the following lines:

> 1. $+$ $(\exists x)Rx$
> 2. $+$ $(\exists x)Gx$
> 3. ? $(\exists x)(Rx \wedge Gx)$
> 4. Ra 1, *EI*
> 5. Ga 2, *EI* (*violates the restriction*)
> 6. $Ra \wedge Ga$ 4, 5, $\wedge I$
> 7. $(\exists x)(Rx \wedge Gx)$ 6, *EG*

The derivation goes off the rails in line 5. In dropping the existential quantifier from the sentence in line 2, we must replace the variable it bound with a constant that does not occur in an earlier active sentence. The constant a occurs in line 4, however. We could drop the existential quantifier in the sentence in line 2 and replace the variable it bound with some constant other than a, with b, for instance, but this would block the conclusion in line 7. The sequence is invalid, so that is precisely what we want to do.

The restriction on rule *EI* requires that the constant introduced when an existential quantifier is dropped not occur in an earlier *active* line of the derivation. Consider the derivation

1.	$+ (\forall x)(Px \supset Qx)$	
2.	$+((\forall x)Px \supset (\exists x)Qx) \supset (\exists x)Rx$	
3.	$? (\exists x)(Rx \vee Sx)$	
4.	$\lceil (\forall x)Px$	
5.	$? (\exists x)Qx$	
6.	Pa	4, EI
7.	$Pa \supset Qa$	1, UI
8.	Qa	6, 7, MP
9.	$\lfloor (\exists x)Gx$	8, EG
10.	$(\forall x)Px \supset (\exists x)Qx$	4–9, CP
11.	$(\exists x)Rx$	2, 10, MP
12.	Ra	11, EI
13.	$Ra \vee Sa$	12, $\vee I$
14.	$(\exists x)(Rx \vee Sx)$	13, EG

Does the introduction of the sentence in line 12 violate the restriction on *EI*? No. Although *a* occurs earlier in the derivation (in lines 6, 7, and 8), the sentences in these lines are not *active*: they occur within matched brackets, that is, within the scope of a supposition that has already been discharged.

Consider one more invalid English sequence

> Someone admires Euterpe.
> Therefore, Euterpe admires herself.

The restriction on rule *EI* blocks the derivation of such sequences. Letting $A①② = $ " ① admires ②" and restricting the domain to persons

1.	$+ (\exists x)Axe$	
2.	$? Aee$	
3.	Aee	1, EI (*violates the restriction*)

The application of *EI* illustrated by the sentence in line 3 violates three clauses in the restriction. First, *e* occurs in an earlier active sentence, namely the sentence in line 1. Second, *e* occurs in the conclusion of the derivation. Third, *e* occurs in ϕ, the expression to which the dropped quantifier is attached. (In line 1, $\phi = Axe$.)

Bear in mind that the restriction on rule *EI* provides a syntactic check on the production of the sorts of invalid derivation illustrated here. In constructing derivations that include applications of *EI*, it is often easiest *first* to drop the existential quantifier and replace the variables it binds with some constant, and *then* to check the resulting sentence to see whether it violates any part of the restriction.

We have seen that rules governing the adding and dropping of quantifiers apply only to sentences occupying whole lines of derivations: a quantifier can be dropped from or added to a sentence only if the sentence to which it is attached is not itself a part of a more inclusive sentence. The sequence below illustrates successive violations of this principle:

$$
\begin{array}{lll}
1. & + & (\exists x)(Fx \wedge Gx) \supset (\exists x)Hx \\
2. & ? & (Fa \wedge Ga) \supset Hb \\
3. & & (Fa \wedge Ga) \supset (\exists x)Hx & \text{2, EI (violates the restriction)} \\
4. & & (Fa \wedge Ga) \supset Hb & \text{3, EI (violates the restriction)}
\end{array}
$$

The quantifiers occurring in the sentence in line 1 include in their scope expressions that are themselves parts of a more inclusive sentence.

Exercises 5.04

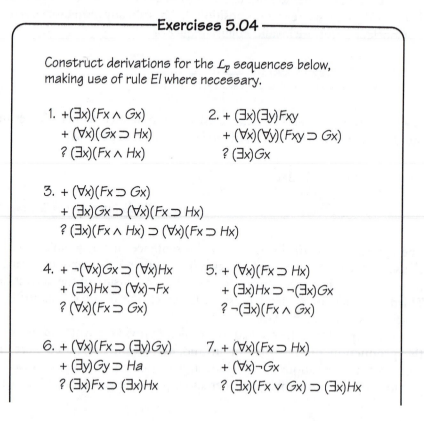

Construct derivations for the L_p sequences below, making use of rule *EI* where necessary.

1. $+(\exists x)(Fx \wedge Gx)$
   $+ (\forall x)(Gx \supset Hx)$
   $? (\exists x)(Fx \wedge Hx)$

2. $+ (\exists x)(\exists y)Fxy$
   $+ (\forall x)(\forall y)(Fxy \supset Gx)$
   $? (\exists x)Gx$

3. $+ (\forall x)(Fx \supset Gx)$
   $+ (\exists x)Gx \supset (\forall x)(Fx \supset Hx)$
   $? (\exists x)(Fx \wedge Hx) \supset (\forall x)(Fx \supset Hx)$

4. $+ \neg(\forall x)Gx \supset (\forall x)Hx$
   $+ (\exists x)Hx \supset (\forall x)\neg Fx$
   $? (\forall x)(Fx \supset Gx)$

5. $+ (\forall x)(Fx \supset Hx)$
   $+ (\exists x)Hx \supset \neg(\exists x)Gx$
   $? \neg(\exists x)(Fx \wedge Gx)$

6. $+ (\forall x)(Fx \supset (\exists y)Gy)$
   $+ (\exists y)Gy \supset Ha$
   $? (\exists x)Fx \supset (\exists x)Hx$

7. $+ (\forall x)(Fx \supset Hx)$
   $+ (\forall x)\neg Gx$
   $? (\exists x)(Fx \vee Gx) \supset (\exists x)Hx$

8. + $(\exists x)(Fx \lor Gx)$
   + $(\forall x)(Fx \supset Hx)$
   + $(\forall x)(Gx \supset Hx)$
   ? $(\exists x)Hx$

9. + $(\exists x)(Fx \land Gx)$
   + $(\exists x)(Hx \land Jx)$
   + $(\exists x)(Fx \land Hx) \supset (\forall x)Kx$
   ? $(\forall x)Fx \supset (\exists x)Kx$

10. + $(\exists x)(\neg Fx \lor Hx)$
    + $(\forall x)(Fx \supset (Gx \supset Hx))$
    ? $(\exists x)(Fx \land Gx) \supset (\exists x)Hx$

11. + $(\exists x)(Fxa \land Hxa)$
    + $(\forall x)(Fxa \supset (Hxa \supset Jx))$
    ? $Fba \supset Jb$

12. + $(\exists x)(Fxa \land Gax)$
    + $(\forall x)(Fxa \supset (\exists y)Hxy)$
    ? $(\exists x)Hxc$

13. + $(\exists x)(Fx \lor Gx)$
    + $(\exists x)Hx \supset (\forall y)\neg Gy$
    ? $(\exists x)(Fx \lor \neg Hx)$

14. + $(\forall x)(Fx \equiv Gx)$
    + $(\forall x)(Fx \supset (Gx \supset Hx))$
    + $(\forall x)Fx \lor (\forall y)Gy$
    ? $(\exists x)Hx$

15. + $(\exists x)(Fx \lor Hx)$
    + $(\exists x)Gx$
    + $(\exists x)Fx \supset (\forall y)(Gy \supset Hy)$
    ? $(\exists x)Hx$

5.05 UNIVERSAL GENERALIZATION (*UG*)

Rule *UI* enables us to drop universal quantifiers and replace the variables bound by those quantifiers with constants. *Universal generalization (UG)* permits us to generalize sentences by adding universal quantifiers and converting individual constants to variables. The rule includes a restriction designed to block inferences of the form

<u>Euterpe is admirable.</u>
Therefore, everyone is admirable.

We are not entitled to infer the conclusion from the premise. It does not follow from Euterpe's being admirable that *everyone* is admirable. We cannot validly derive a universal conclusion from a premise that concerns a particular named object.

The restriction blocks, as well, inferences of the form

<u>Something is admirable.</u>
Therefore, everything is admirable.

From the fact that *F* is true of something, it does not follow that *F* is true of everything.

Let us first set out rule *UG*, together with its restriction, and then see how it applies in a particular case

$$(UG) \quad \phi\alpha/\beta \vdash (\forall\alpha)\phi$$

Restriction: β does not occur in a premise, in an undischarged supposition, or in ϕ; nor does β occur in an active sentence as a result of an application of *EI*. (A supposition is discharged when it occurs within matched brackets. A sentence is active when it is neither preceded by a ? nor occurs within matched brackets.)

Rule *UG* is founded on the idea that if we can show that something holds of an arbitrary object in a collection of objects, we have shown that it holds for every object in the collection. In proving the Pythagorean theorem (the square of the hypotenuse of a right triangle = the sum of the squares of the remaining sides), we might begin by considering an arbitrary right triangle with sides *a*, *b*, and *c*. In proving that $a^2 = b^2 + c^2$, we in effect prove that this holds for any right triangle. The restriction associated with *UG* is designed to insure that when a universal quantifier is added to a sentence and the variables it binds are converted from individual constants, those individual constants function in the derivation as the names of an arbitrary object. The first part of the restriction is obvious: if an individual constant occurs in a premise of a derivation, it does not function as a designation for an arbitrary object. We can come to understand the restriction fully by considering its application in derivations.

Let us begin with the sequence

All horses are quadrapeds.
<u>All quadrapeds have spleens.</u>
Therefore, all horses have spleens.

The sequence is valid. We can prove that it is valid in $\mathcal{L}_p$ by means of an application of *UG*. We can suppose that $Hx =$ "*x* is a horse," $Qx =$ "*x* is a quadraped," and $Sx =$ "*x* has a spleen."

1. $+ (\forall x)(Hx \supset Qx)$
2. $+ (\forall x)(Qx \supset Sx)$

   3. ? $(\forall x)(Hx \supset Sx)$

   4. $Ha \supset Qa$ 1, UI

   5. $Qa \supset Sa$ 2, UI

   6. $Ha \supset Sa$ 4, 5, HS

   7. $(\forall x)(Hx \supset Sx)$ 5, UG

The sentence in line 4 results from reasoning as follows: if all horses are quadrapeds, then if an arbitrary object, a, is a horse, a is a quadraped. Similarly, the sentence in line 5 is obtained by reasoning that if all quadrapeds have spleens, then if an arbitrary object, a, is a quadraped, a has a spleen. The sentence in line 6 results from reasoning in accord with HS: if it is true that if a is a horse, a is a quadraped, and true as well that if a is a quadraped, a has a spleen, then it is true that if a is a horse, a has a spleen. So long as a represents a genuinely arbitrary object, what holds for a must hold for any object; hence the sentence in line 7.

The application of UG in the derivation accords with the restriction. The individual constant does not occur in a premise. Nor does a occur in an undischarged supposition; there are no suppositions, so there are no undischarged suppositions. Finally, a does not occur in $(Hx \supset Sx)$, the expression to which the quantifier is affixed; there are no applications of rule EI, so the last clause in the restriction is satisfied.

The sense in which individual constants function as names of arbitrary *objects* should not be confused with the sense in which constants function as arbitrary *names* in applications of rule EI. If we know that something is F (i.e., $(\exists x)Fx$), then we can designate an arbitrary name to stand for that something. In applying rule UG, in contrast, we reason that if F is true of an arbitrary object, an object arbitrarily selected from a class of objects, F is true of every object in the class. If something holds true of an arbitrary right triangle, or an arbitrary tiger, it holds true of *any* triangle or tiger.

We began the section with examples of English sequences that were clearly invalid. Let us look more closely at the first of those sequences.

> <u>Euterpe is admirable.</u>
> Therefore, everyone is admirable.

We can translate the sequence into $\mathcal{L}_p$ (letting $Ax = $ "x is admirable")

   1. + Ae

   2. ? $(\forall x)Ax$

   3. $(\forall x)Ax$ 1, UG (*violates the restriction*)

The application of *UG* violates the restriction. We cannot convert an individual constant to a variable that is picked up by a universal quantifier if that constant occurs in a premise. In this case, *e*, the individual constant we are converting to a variable, appears in the premise in line 1.

The second example violates the restriction in a different way

> Something is admirable.
> Therefore, everything is admirable.

In $\mathcal{L}_p$

1.	+	$(\exists x)Ae$	
2.	?	$(\forall x)Ax$	
3.		Aa	1, *EI*
4.		$(\forall x)Ax$	3, *UG* (*violates the restriction*)

The restriction on *UG* is expressed syntactically, but its aim is to block invalid inferences of the sort illustrated by this sequence. In this case, the inference is blocked by not permitting universal quantifiers to be added in such a way that the individual constant converted to a variable and picked up by the new quantifier—in this case *a*—was not introduced by an application of rule *EI*. In the example above, *a was* introduced in line 3 by an application of *EI*, so the inference is blocked.

Another way in which the restriction on *UG* can be violated is illustrated by the invalid English sequence

> Euterpe admires herself.
> Therefore, everyone admires Euterpe.

Suppose we attempt to construct a derivation for this sequence in $\mathcal{L}_p$, again letting $A①② =$ "① admires ②"

1.	+	Aee	
2.	?	$(\forall x)Axe$	
3.		$(\forall x)Axe$	1, *UG* (*violates the restriction*)

The restriction on *UG* blocks inferences of this sort by requiring that ϕ, the expression to which the added universal quantifier is affixed, not include instances of β, the constant being converted to a variable and picked up by the quantifier. The restriction amounts to the requirement that in an application of *UG*, we can convert an instance of an individual constant to a variable picked up by an added universal quantifier only if we convert *every* instance of that individual constant to the variable. (You will have noticed that the application of rule *UG* in the derivation above violates the restriction in another respect as well. The constant *e* occurs in a premise. As a result, in the context of this derivation, *e* cannot be used to designate an arbitrary object.)

An illustration of the third clause in the restriction on *UG* requires that we look at a proof involving a supposition

> <u>Everything with a mane is a quadraped.</u>
> Therefore, horses with manes are quadrapeds.

Using the interpretation of *Hx* and *Qx* set out above, and letting *Mx* = "*x* has a mane," we can construct a derivation for this sequence using *CP*

1. +	$(\forall x)(Mx \supset Qx)$	
2. ?	$(\forall x)((Hx \land Mx) \supset Qx)$	
3.	$Ma \supset Qa$	1, UI
4.	⌈ $Ha \land Ma$	
5.	| ? Qa	
6.	| Ma	4, $\land E$
7.	⌊ Qa	3, 6, MP
8.	$(Ha \land Ma) \supset Qa$	4–7, CP
9.	$(\forall x)((Hx \land Mx) \supset Qx)$	8, UG

In adding a universal quantifier in line 9, we convert an individual constant, *a*, to a variable. Have we violated the restriction on *UG*? The premise in line 1 does not contain an occurrence of *a*, and, in adding the quantifier in line 9, we have converted every instance of *a* to an *x* that is picked up by the quantifier. There are no applications of rule *EI*, so we have not violated the clause in the restriction that blocks the conversion of a constant added via applications of *EI* to a variable that is then picked up by a new universal quantifier.

What about the remaining clause in the restriction? Does *a* occur in an undischarged supposition? The derivation includes a supposition (the sentence in line 4), and that supposition contains instances of *a*. The universal quantifier is added to the sentence in line 8; however, a sentence that occurs *outside* the scope of the supposition, that is, after the supposition has been discharged.

A supposition is discharged once ∟ has been entered and the suppositional portion of a derivation is "closed off." Sentences included within the matched brackets become inactive, unavailable for further inferences. Perhaps this can be made clearer by considering an invalid sequence and a derivation in which the restriction we have been exploring is violated. The sequence below is invalid:

All horses are quadrapeds.
All quadrapeds have spleens.
Therefore, if Ed is a horse, everything has a spleen.

Its invalidity is reflected in L_p by its violation of the restriction on UG concerning the occurrence of individual constants in sentences introduced as suppositions

1.	+ $(\forall x)(Hx \supset Qx)$	
2.	+ $(\forall x)(Qx \supset Sx)$	
3.	? $He \supset (\forall x)Sx$	
4.	⌈ He	
5.	│ ? $(\forall x)Sx$	
6.	│ $He \supset Qe$	1, UG
7.	│ $Qe \supset Se$	2, UG
8.	│ $He \supset Se$	6, 7, HS
9.	│ Se	4, 8, HS
10.	∟ $(\forall x)Sx$	8, UG (*violates the restriction*)
11.	$He \supset (\forall x)Sx$	4–10, CP

In the sentence in line 10, an individual constant, e, has been converted to a variable, x, and a universal quantifier has been added to pick up that variable. This violates the restriction on UG. The rule has been applied *within* the scope of the sentence introduced in line 4 as a supposition; that supposition both contains e and has not yet been discharged.

The restriction on UG does not prohibit the addition of a universal quantifier within the scope of a supposition. The application of UG in line 10 is incorrect, not because it occurs within the scope of a supposition, but because it occurs within the scope of a supposition that includes e, the individual constant that will be converted to a variable and picked up by the universal quantifier ($\forall x$).

Let us consider one more derivation featuring an application of rule UG. Consider the sequence

All pigs have spleens.
Therefore, all pigs have spleens or livers.

It is easy to prove the sequence valid using *CP*. However, the derivation below is incorrect:

1. + $(\forall x)(Px \supset Sx)$
2. ? $(\forall x)(Px \supset (Sx \vee Lx))$
3. ⌈ Px (*not a sentence!*)
4.   ? $Sx \vee Lx$
5. $Px \supset Sx$ 1, UI (*not a sentence!*)
6. Sx 3, 5, MP (*not a sentence!*)
7. ⌊ $Sx \vee Lx$ 6, $\vee$I (*not a sentence!*)
8. $Px \supset (Sx \vee Lx)$ 3–7, CP (*not a sentence!*)
9. $(\forall x)(Px \supset (Sx \vee Lx))$ 8, UG

Every line of a derivation must consist of a sentence. The expression on line 3, and those on lines 5–8 all contain free variables. As a result, these expressions are not sentences of $\mathcal{L}_p$. The error is avoidable, as the derivation below illustrates

1. + $(\forall x)(Px \supset Sx)$
2. ? $(\forall x)(Px \supset (Sx \vee Lx))$
3. ⌈ Pa
4.   ? $Sa \vee La$
5. $Pa \supset Sa$ 1, UI
6. Sa 3, 5, MP
7. ⌊ $Sa \vee La$ 6, $\vee$I
8. $Pa \supset (Sa \vee La)$ 3–7, CP
9. $(\forall x)(Px \supset (Sx \vee Lx))$ 8, UG

In this derivation, every line consists of a sentence of $\mathcal{L}_p$.

Exercises 5.05

Construct derivations for the L_p sequences below, using rule UG where necessary.

1. $+ (\forall x)(Fx \supset Gx)$
   $+ \neg(\exists x)(Gx \wedge Hx)$
   $? (\forall x)(Fx \supset \neg Hx)$

2. $+ (\forall x)(Fx \supset Gx)$
   $+ (\exists x)Gx \supset (\forall x)(Fx \supset Hx)$
   $? (\forall x)Fx \supset (\forall x)Hx$

3. $+ (\forall x)(Fx \supset Gx)$
   $? (\forall x)((\exists y)(Fy \wedge Hxy) \supset (\exists y)(Gy \wedge Hxy))$

4. $+ (\forall x)(Fx \wedge Gx)$
   $? (\forall x)Fx \wedge (\forall x)Gx$

5. $+ Fa \supset (\forall x)Gax$
   $? (\forall x)(Fa \supset Gax)$

6. $+ (\exists x)(Fx \vee Hx)$
   $+ (\forall x)(Hx \supset Fx)$
   $+ (\exists x)Gx \supset (\forall x)(Gx \supset Hx)$
   $? (\forall x)(Gx \supset Fx)$

7. $+ (\forall x)(Fx \supset Gx)$
   $+ (\forall x)(Fx \supset Hx)$
   $? (\forall x)(Fx \supset (Gx \wedge Hx))$

8. $+ (\forall x)(Fx \supset Gx)$
   $+ (\forall x)(\exists y)(Fy \wedge Hxy)$
   $? (\forall x)(\exists y)(Gy \wedge Hxy)$

9. $+ (\forall x)Fx$
   $+ (\forall x)\neg Gx$
   $? \neg(\exists x)(Fx \equiv Gx)$

10. $+ (\forall x)(\forall y)(Fxy \supset (\exists z)Gzxy)$
    $+ (\forall x)(\forall y)(\forall z)(Gzxy \supset (Hzx \wedge Hzy))$
    $+ (\forall x)Fxa$
    $? (\forall x)(\exists y)(Gyxa \supset Hya)$

11. $+ (\exists x)(Fx \wedge (\forall y)(Gy \supset Hxy))$
    $+ (\forall x)(Fx \supset (\forall y)(Jy \supset \neg Hxy))$
    $? (\forall x)(Gx \supset \neg Jx)$

12. $+ (\exists x)(\exists y)Fxy$
    $+ (\forall x)((\exists y)Fxy \supset (\exists y)((\forall z)Fyz \wedge Fxy))$
    $? (\exists x)(\forall y)Fxy$

13. $+ (\forall x)((\exists y)Gyx \supset Gxx)$
    $+ (\forall x)(Fx \supset ((\exists y)Gxy \supset (\exists y)Gyx))$
    $+ \neg(\forall x)Gxx$
    $? (\forall x)(Fx \supset (\forall y)\neg Gxy)$

14. $+ (\forall x)(\forall y)((Fax \land Fya) \supset Fxy)$
    $+ (\forall x)(Gx \supset Fxa)$
    $+ (\exists x)(Gx \land Fax)$
    $? (\exists x)(Gx \land (\forall y)(Gy \supset Fxy))$

15. $+ (\forall x)((Fx \land \neg(\exists y)Hxy) \supset Gx)$
    $+ (\forall x)(Jx \supset (Fx \land \neg(Kx \lor Gx)))$
    $? (\forall x)(Jx \supset (\exists y)Hxy)$

5.06 IDENTITY (*ID*)

Identity, as we saw in § 4.09, is a relation every object bears to itself and to no other object. Identity statements, like "Bruce Wayne is Bruce Wayne" are trivial, but not all identity statements are trivial. Objects can be designated by more than one singular term. A sentence like

Batman is (i.e., *is identical with*) Bruce Wayne

is true if and only if the object "Batman" purports to designate is the very same object as the object "Bruce Wayne" purports to designate. Similarly

The author of *Jabberwocky* is Dodgson

is true just in case "the author of *Jabberwocky*" and "Dodgson" codesignate. Consider an English sequence that includes a statement of identity

Carroll is an author and a don.
Carroll is Dodgson.
Therefore, Dodgson is an author and a don.

The sequence is valid. We can represent it in $\mathcal{L}_p$ as follows (assuming that Ax = "x is an author" and Dx = "x is a don):

    1. + $Ac \wedge Dc$
    2. + $c = d$
    3. ? $Ad \wedge Dd$

To derive the conclusion from the premises in lines 1 and 2, we need a rule that will enable us to make use of sentences expressing identities. The rule includes two parts, each part reflecting an aspect of the concept of identity. In the formulation of the rule, β and γ stand for individual constants, and ϕ and ψ designate sentences that are the same except that ϕ includes instances of β where ψ includes instances of γ

    (ID)   (i) $\vdash \beta = \beta$
          (ii) $\phi, \beta = \gamma \vdash \psi\beta/\gamma$

    The first clause of ID permits us to enter on any line of a derivation a sentence consisting of an identity sign flanked by two occurrences of a single individual constant. This reflects the logical truth that every object is identical with itself.

    ID's second clause permits us to infer from a sentence containing an individual name, β, together with a statement of identity, $\beta = \gamma$, to another sentence that differs from the first only in its substitution of γ for β. This reflects our conviction that if a and b are identical, then whatever is true of a is true of b, and *vice versa*. If Batman is lefthanded and has a scar on his forearm then, if Batman is identical with Bruce Wayne, Bruce Wayne is lefthanded and has a scar on his forearm.

    This pattern of reasoning is evident in the sequence introduced above. If Carroll is an author and a don, then if Carroll and Dodgson are one and the same person, Dodgson is an author and a don

    1. + $Ac \wedge Dc$
    2. + $c = d$
    3. ? $Ad \wedge Dd$
    4. $Ad \wedge Dd$ 1, 2, ID

Suppose that we discover that Carroll had brown hair and a limp, but that Dodgson did not. It would follow that Carrol and Dodgson are not one and the same. We can show this in $\mathcal{L}_p$ (letting $Bx = $ "x has brown hair" and $Lx = $ "x limps")

1. +	$Bc \wedge Lc$	
2. +	$\neg(Bd \wedge Ld)$	
3. ?	$c \neq d$	
4.	$\lceil c = d$	
5.	$? \times$	
6.	$Bd \wedge Ld$	1, 4, ID
7.	$\lfloor (Bd \wedge Ld) \wedge \neg(Bd \wedge Ld)$	2, 6 $\wedge I$
8.	$c \neq d$	4–7, IP

When might we put the first clause of rule *ID* to work? Consider the sequence

<u>Carroll is an author.</u>
Therefore, something is an author and identical with Carroll.

This sequence can be represented in $\mathcal{L}_p$ and proved valid as follows:

1. +	Ac	
2. ?	$(\exists x)(Ax \wedge x = c)$	
3.	$\lceil \neg(\exists x)(Ax \wedge x = c)$	
4.	$? \times$	
5.	$(\forall x)\neg(Ax \wedge x = c)$	3, QT
6.	$\neg(Ac \wedge c = c)$	5, UI
7.	$c \neq c \vee \neg Ac$	6, DeM
8.	$c \neq c$	1, 7, $\vee E$
9.	$c = c$	ID
10.	$\lfloor c = c \wedge c \neq c$	8, 9. $\wedge I$
11.	$(\exists x)(Ax \wedge x = c)$	3–10, IP

The sentence introduced via an application of *ID* in line 9 is not derived from any earlier sentence, so no line numbers appear in its justification.

Let us look at one more derivation involving identity. Consider the English sequence

Someone with a gun shot Elvis.
<u>Only Fenton and Gertrude have guns.</u>
Therefore, either Fenton or Gertrude shot Elvis.

Let us restrict the domain to persons and let $Gx = $ "x has a gun" and $S\text{①②} = $ "① shot ②." How might we represent this sequence in $\mathcal{L}_p$? The first sentence and the conclusion are unproblematic. The second premise is more challenging. We can make progress by translating the second premise into pidgin L_p

> For all x, if x has a gun, then x is identical with
> Fenton or x is identical with Gertrude.

The sequence can now be translated into L_p

1. $+$ $(\exists x)(Gx \wedge Sxe)$
2. $+$ $(\forall x)(Gx \supset (x = f \vee x = g))$
8. $?$ $Sfe \vee Sge$

and the conclusion derived from the premises

1.	$+$	$(\exists x)(Gx \wedge Sxe)$	
2.	$+$	$(\forall x)(Gx \supset (x = f \vee x = g))$	
8.	$?$	$Sfe \vee Sge$	
4.		$\neg(Sfe \vee Sge)$	
5.		$? \times$	
6.		$Ga \wedge Sae$	1, EI
7.		Ga	6, $\wedge E$
8.		Sae	6, $\wedge E$
9.		$Ga \supset (a = f \vee a = g)$	2, UI
10.		$a = f \vee a = g$	7, 9, MP
11.		$a = f$	
12.		$? Sfe$	
13.		Sfe	8, 11, ID
14.		$a = f \supset Sfe$	11–13, CP
15.		$a = g$	
16.		$? Sge$	
17.		Sge	8, 11, ID
18.		$a = g \supset Sge$	15–17, CP
19.		$Sfe \vee Sge$	10, 14, 18, CD
20.		$(Sfe \vee Sge) \wedge \neg(Sfe \vee Sge)$	4, 19 $\wedge I$
21.		$Sfe \vee Sge$	4–20, IP

The derivation is worth working through. It includes an application of rule CD (in line 19) set up by a pair of embedded applications of CP (in lines 11–14 and 15–18).

─────Exercises 5.06─────

Construct derivations for the $\mathcal{L}_p$ sequences below,
making use of rule ID where necessary.

1. + $(\forall x)(x = a \supset Fx)$
   + $(\forall x)(Fx \supset Fb)$
   ? Fb

2. + $(\exists x)(Fx \wedge Gx)$
   + $(\exists x)(Fx \wedge \neg Gx)$
   ? $(\exists x)(\exists y)((Fx \wedge Fy) \wedge x \neq y)$

3. + $(\forall x)(\exists y)Fxy$
   + $(\forall x)\neg Fxx$
   ? $(\exists x)(\exists y)(x \neq y)$

4. + $(\forall x)(Fx \supset (\forall y)(Fy \supset x = y))$
   + $(\exists x)(Fx \wedge Gx)$
   ? $(\forall x)(Fx \supset Gx)$

5. + $(\exists x)(\forall y)(Fy \equiv x = y)$
   + $Fa \wedge Fb$
   ? $a = b$

6. + $(\exists x)(Fx \wedge (\forall y)(Fy \supset x = y))$
   + $\neg Fb$
   ? $(\exists x)(x \neq b)$

7. + $(\forall x)(Fx \supset Gx)$
   + $(\forall x)(Gx \supset Hx)$
   + $Fa \wedge \neg Hb$
   ? $a \neq b$

8. + $(\exists x)((Fx \wedge Gax) \wedge Hx)$
   + $Fb \wedge Gab$
   + $(\forall x)((Fx \wedge Gax) \supset x = b)$
   ? Hb

9. + $(\exists x)((Fx \wedge (\forall y)(Fy \supset x = y)) \wedge Gx)$
   + $\neg Ga$
   ? $\neg Fa$

10. + $(\forall x)(\forall y)((Fxy \wedge x \neq y) \supset Gxy)$
    + $(\exists x)(\forall y)(x \neq y \supset Fxy)$
    ? $(\exists x)(\forall y)(x \neq y \supset Gxy)$

11. + $\neg(\exists x)(Fx \wedge (\forall y)(Fy \supset x = y))$
    + $(\exists x)Fx$
    ? $(\exists x)(\exists y)((Fx \wedge Fy) \supset x \neq y)$

12. + $(\exists x)(Fx \wedge (\forall y)(Fy \supset x = y))$
    + $(\exists x)(Fx \wedge Gx)$
    ? $(\forall x)(Fx \supset Gx)$

13. $+ \neg(\exists x)(Fx \wedge (\forall y)(Fy \supset x = y))$
    $+ (\exists x)Fx$
    $? (\exists x)(\exists y)((Fx \wedge Fy) \wedge x \neq y)$

14. $+ (\forall x)(\exists y)Fxy$
    $+ \neg(\exists x)Fxx$
    $? (\forall x)(Fxa \supset a \neq x)$

15. $+ (\exists x)(x = a \wedge x = b)$
    $+ Fa$
    $? Fb$

5.07 THEOREMS OF $\mathcal{L}_p$

Theorems of $\mathcal{L}_p$ express logical truths. We can establish that a sentence is a theorem of $\mathcal{L}_p$ by deriving it from the empty set of sentences. The technique for deriving theorems resembles the technique employed to derive theorems of $\mathcal{L}_s$. The only difference is that we now must contend with quantifiers.

Consider the theorem

$$\vdash (\exists x)(Fx \wedge Gx) \supset ((\exists x)Fx \wedge (\exists x)Gx)$$

We can show that this sentence is a theorem by deriving it, via CP, from the empty set of sentences

1.	$(\exists x)(Fx \wedge Gx)$	
2.	$? (\exists x)Fx \wedge (\exists x)Gx$	
3.	$Fa \wedge Ga$	1, EI
4.	Fa	3, ∧E
5.	Ga	3, ∧E
6.	$(\exists x)Fx$	4, EG
7.	$(\exists x)Gx$	5, EG
8.	$(\exists x)Fx \wedge (\exists x)Gx$	6, 7, ∧I
9.	$(\exists x)(Fx \wedge Gx) \supset ((\exists x)Fx \wedge (\exists x)Gx)$	1–8, CP

The proof is straightforward. Proofs of other theorems can be more challenging. Consider the theorem

$$\vdash (\forall x)(\exists y)(Fx \wedge Gy) \supset (\exists x)(Fx \wedge Gx)$$

We can construct a derivation of this theorem using *CP*

1. $\quad$ $(\forall x)(\exists y)(Fx \wedge Gy)$
2. $\quad$? $(\exists x)(Fx \wedge Gx)$
3. $\quad$ $(\exists y)(Fa \wedge Gy)$ $\qquad\qquad\qquad$ 1, *UI*
4. $\quad$ $Fa \wedge Gb$ $\qquad\qquad\qquad\qquad$ 3, *EI*
5. $\quad$ $(\forall x)(Fx \wedge Gb)$ $\qquad\qquad\quad$ 4, *UG*
6. $\quad$ $Fb \wedge Gb$ $\qquad\qquad\qquad\qquad$ 5, *UI*
7. $\quad$ $(\exists x)(Fx \wedge Gx)$ $\qquad\qquad\quad$ 6, *EG*
8. $(\forall x)(\exists y)(Fx \wedge Gy) \supset (\exists x)(Fx \wedge Gx)$ $\qquad$ 1–7, *CP*

You may be suspicious of the application of rule *UG* in line 5, but that application of the rule does not violate the restriction on *UG*; *a* does not occur in the suppositional premise in line 1, nor is "*a*" introduced in a sentence obtained by an application of rule *EI*.

Derivations of theorems often incorporate moves of the sort illustrated above; these moves are so simple they sometimes do not occur to us. The derivation of the theorem below illustrates another feature common to derivations of theorems

$$\vdash (\exists x)(Fx \supset Ga) \equiv ((\forall x)Fx \supset Ga)$$

This theorem consists of a biconditional sentence. Bearing in mind that a biconditional is a two-way conditional, we can construct a derivation by deploying a pair of conditional proofs

1. $\quad$ $(\exists x)(Fx \supset Ga)$
2. $\quad$? $(\forall x)Fx \supset Ga$
3. $\quad\quad$ $(\forall x)Fx$
4. $\quad\quad$?Ga
5. $\quad\quad$ $Fb \supset Ga$ $\qquad\qquad\qquad$ 1, *EI*
6. $\quad\quad$ Fb $\qquad\qquad\qquad\qquad\quad$ 3, *UI*
7. $\quad\quad$ Ga $\qquad\qquad\qquad\qquad\quad$ 5, 6, *MP*
8. $\quad$ $(\forall x)Fx \supset Ga$ $\qquad\qquad$ 3–7, *CP*
9. $(\exists x)(Fx \supset Ga) \supset ((\forall x)Fx \supset Ga)$ $\qquad$ 1–8, *CP*
10. $\quad$ $(\forall x)Fx \supset Ga$
11. $\quad$? $(\exists x)(Fx \supset Ga)$
12. $\quad\quad$ $\neg(\exists x)(Fx \supset Ga)$
13. $\quad\quad$? $\times$
14. $\quad\quad$ $(\forall x)\neg(Fx \supset Ga)$ $\qquad\qquad$ 12, *QT*

15.	$\neg(Fb \supset Gb)$	14, UI
16.	$\neg(\neg Fb \vee Gb)$	15, Cond
17.	$Fb \wedge \neg Gb$	16, DeM
18.	Fb	17, $\wedge E$
19.	$(\forall x)Fx$	18, UG
20.	Ga	10, 19, MP
21.	$\neg Ga$	17, $\wedge E$
22.	$Ga \wedge \neg Ga$	20, 21, $\wedge I$
23.	$(\exists x)(Fx \supset Ga)$	12–22, CP
24.	$((\forall x)Fx \supset Ga) \supset (\exists x)(Fx \supset Ga)$	10–23, CP
25.	$((\exists x)(Fx \supset Ga) \supset ((\forall x)Fx \supset Ga)) \wedge$ $\quad (((\forall x)Fx \supset Ga) \supset (\exists x)(Fx \supset Ga))$	9, 24, $\wedge I$
26.	$(\exists x)(Fx \supset Ga) \equiv ((\forall x)Fx \supset Ga)$	25, Bicond

In line 25, the conditionals appearing in lines 9 and 24 are conjoined so that in line 26, *Bicond* can be applied yielding the theorem. Note that when the existential quantifier is dropped in line 5, the variable it bound cannot be replaced by *a*: *a* occurs in an earlier active sentence, namely the sentence in line 1. The application of *UG* in line 19 may look suspicious, but it accords with the restriction: *b*, the constant figuring in the generalization, does not occur in an active supposition or in the conclusion, nor does it result from an application of rule *EI*.

Exercises 5.07

Provide derivations for each of the theorems of $\mathcal{L}_p$ below.

1. $\vdash (\forall x)(\exists y)(Fy \supset Fx)$

2. $\vdash (\forall x)(Fx \vee \neg Fx)$

3. $\vdash \neg(\exists x)(Fx \wedge \neg Fx)$

4. $\vdash (\forall x)(\forall y)((Fx \wedge x = y) \supset Fy)$

5. $\vdash (\forall x)(\forall y)(x = y \supset y = x)$

6. $\vdash (\forall x)(\forall y)(x = y \supset (Fx \equiv Fy))$

7. ⊢ $(\exists x)(\forall y)(Fx \supset Fy)$

8. ⊢ $(\exists x)(\forall y)Fxy \supset (\forall y)(\exists x)Fxy$

9. ⊢ $(\exists x)Fx \supset ((\forall x)\neg Fx \supset (\forall x)Gx)$

10. ⊢ $(\forall x)(Fx \supset Gx) \supset (\neg(\exists x)Gx \supset \neg (\exists x)Fx)$

11. ⊢ $(\exists x)Fxx \supset (\exists x)(\exists y)Fxy$

12. ⊢ $(\forall x)(\forall y)Fxy \supset (\forall x)Fxx$

13. ⊢ $((\exists x)Fx \wedge (\exists x)Gx) \supset (\exists x)(\exists y)(Fx \wedge Gy)$

14. ⊢ $(\forall x)(\forall y)(Fxy \supset \neg Fxy) \supset (\forall x)\neg Fxx$

15. ⊢ $(\exists x)(\forall y)(Fx \supset Gy) \supset ((\forall x)Fx \supset (\forall x)Gx)$

5.08 INVALIDITY IN L_p

If the rules associated with L_p have been properly formulated, they will yield only valid inferences. Imagine that you have been unsuccessful in discovering a derivation that proves a sequence valid. There are two possibilities. First, a derivation may be possible, but it may escape your ingenuity. Second, the sequence may be *invalid*. Were that so, a derivation establishing its validity would not be possible. You could settle the matter decisively by discovering a derivation showing that the sequence *is* valid. Failure to discover a derivation, however, would not entitle you to conclude that the sequence is not valid.

Might there be a *test* for invalidity, a procedure that would enable us to prove particular sequences invalid? In L_s, the matter was simple. We could always construct a truth table—or a chart—that demonstrated invalidity conclusively. Truth tables do not extend smoothly to L_p. Truth tables are finite devices, but the quantificational structure of L_p is *open-ended*. Consider a universally quantified sentence of the form $(\forall x)Fx$: everything is

F. This sentence is true under an interpretation, if and only if every object in the domain is *F*. This means that we can regard universally quantified sentences as *reducible* to conjunctions. If $(\forall x)Fx$, then *a* is *F*, *b* is *F*, *c* is *F*, *and so on*. This is fine so long as the domain is finite. In that case, $(\forall x)Fx$ is reducible to a finite, although perhaps very long, conjunction. Suppose the domain contains an infinite number of objects, however. In that case $(\forall x)Fx$ reduces to an *infinite* conjunction

$$Fa \land Fb \land Fc \land \dots$$

Truth tables are finite, however, so such a conjunction could not be represented, much less tested, by means of a truth table.

In one respect it is misleading to describe universally quantified sentences as reducible to conjunctions. Sentences must be finite in length. If we consider infinite domains, however, conjunctive reductions of universally quantified sentences would be infinitely long. Such conjunctions would not count as sentences; hence they would not count as sentences reductively equivalent to the original universally quantified sentence. We can finesse this difficulty by supposing that universally quantified sentences are reducible to conjunctions, even though some conjunctions, namely those infinite in length, do not count as sentences of L_p.

Parallel remarks apply to existentially quantified sentences. If a universally quantified sentence can be regarded as reducible to a (possibly infinite) conjunction, an existentially quantified sentence of the form $(\exists x)Fx$ can be regarded as reducible to a (possibly infinite) *disjunction*. If something is *F*, then *a* is *F*, or *b* is *F*, or *c* is *F*, *and so on*.

Now imagine setting out to test a sequence in L_p for validity by means of a truth table. Truth tables provide a systematic representation of the truth conditions of sentences in the sequence. Our aim is to discover whether, given those truth conditions, it is possible for the sequence's premises to be true when its conclusion is false. In truth table lingo, this means discovering whether a truth table representation of the sequence contains a row in which the value of each premise is *true* and the value of the conclusion is *false*. As a first step, we could replace quantified sentences with conjunctive and disjunctive reductions. How many conjuncts and disjuncts? At this point simple comparisons between L_s and L_p break down. The number of conjuncts and disjuncts depends on the size of the domain. If the domain is infinite, then quantified sentences will require infinitely long counterparts, and the truth table method of checking for validity goes by the board.

Let us back up and consider what is required for the validity of a sequence, $\langle P,c \rangle$. A sequence, $\langle P,c \rangle$, is valid just in case P logically implies c,

($P \vDash c$). We know that P logically implies c just in case there is no interpretation under which P is true and c is false. What is involved in giving an interpretation of sentences in L_p? An interpretation includes the specification of a domain of objects and an assignment of constants to objects in the domain (see § 4.14 for a more detailed discussion). So long as we allow for the possibility of infinite domains, we cannot avail ourselves of the truth table method of checking the validity of sequences in L_p containing quantified sentences. There is no reason to restrict the choice of domains, however, so we cannot rely on truth tables to provide foolproof tests for validity in L_p.

We can, however, make use of truth tables as limited aids—heuristics— for establishing the invalidity of sequences in L_p. A sequence is invalid if there is an interpretation—*any* interpretation—under which its premises are true and its conclusion false. Interpretations in L_p include the specification of a domain. Suppose we specify an interpretation with a finite domain for a sequence, $\langle P,c \rangle$; and suppose, under that interpretation, P is true and c is false. We could do this by replacing universally quantified sentences in the sequence with finite conjunctions, replacing existentially quantified sentences with finite disjunctions, and providing a truth table representation of the sequence in which, in at least one row, P is true and c is false. How might this work in practice?

Consider the sequence

    1. $+$ $(\forall x)(Fx \supset Gx)$
    2. $+$ $\neg(\exists x)Fx$
    8. ? $\neg(\exists x)Gx$

Let us provide an interpretation of the sentences in the sequence using a domain consisting of a single object, a. Given this domain, the sentence in line 1 is reducible to

$$Fa \supset Ga$$

If everything in the domain is such that if it is an F, then it is a G, and if the domain consists of just one object, a, then it is true of a that if a is F then a is G. The sentence in line two says that it is not the case that something is F. There is only a single object in the domain to consider, namely a, so this sentence is reducible to

$$\neg Fa$$

In pidgin L_p: "It's not the case that a is F." Finally, the conclusion can be rewritten as

$$\neg Ga$$

If it is not the case that something is G, and if we have only one object to consider, a, then it is not the case that a is G.

We can now represent the sequence by means of a truth table

Fa Ga	Fa ⊃ Ga	¬Fa	¬Ga
T T	T	F	F
T F	F	F	T
F T	T	T	F
F F	T	T	T

We can see that in the third row of the truth table, the premises are true and the conclusion is false: the sequence is invalid.

In providing an interpretation we begin by specifying a domain of objects. Each row of the truth table above represents one member of a class of interpretations for that domain. So long as the domain is finite, we can exhaustively list the possible interpretations for a (finite) sequence of sentences. We can now see that so long as we restrict the domain to a single individual, differences between quantifiers are invisible. Thus, in a domain consisting of a single object, a, the sentence

$$(\forall x)Fx$$

and the sentence

$$(\exists x)Fx$$

are reducible to identical sentences. If every object is F and there is only one object, a, then a is F

$$Fa$$

Similarly, if at least one object is F and there is only a single object, a, then a is F

$$Fa$$

Fermat's Last Theorem

If you have ever struggled with a sequence for which you can find no derivation, you can imagine the frustration felt by mathematicians who, for more than three hundred years, could not find a proof for the theorem:

The equation $x^n + y^n = z^n$, where n is an integer greater than 2, has no solution in the positive integers.

The theorem, commonly known as "Fermat's Last Theorem," is named after French mathematician, Pierre de Fermat (1601–1665), who wrote in the margin of a book "I have found a truly wonderful proof which this margin is too small to contain."

In 1993, Andrew Wiles, a British mathematician working in the United States, announced that he had discovered a proof of the theorem. The proof is complex and involves subproofs that only a handful of mathematicians are capable of evaluating. As a result, it is not yet clear whether Wiles's proof will stand up. Although there are indications that it will, if it does not, it will not be the first time an apparent proof of Fermat's theorem has been shown to be flawed.

Although you may suspect that some of the sequences for which you are asked to provide derivations are the logical counterparts of Fermat's Last Theorem, rest assured that derivations required for the exercises in this book are all a breeze by comparison.

Interpretations that make use of domains consisting of a single object cannot distinguish universally and existentially quantified sentences. Matters change once we move to interpretations incorporating domains consisting of two or more objects. Consider the sentences above construed for a domain consisting of two objects, a and b. The universally quantified sentence is reducible to a conjunction

$$Fa \land Fb$$

If every object in the domain is F and the domain contains exactly two objects, a and b, then a is F *and* b is F. In contrast, if, as in the existentially quantified version of the sentence, *some* object is F, then a is F *or* b is F

$$Fa \lor Fb$$

This brings us back to the notion that universally quantified sentences are reducible to conjunctions and existentially quantified sentences are reducible to disjunctions.

Now consider the sequence

      1. $+$ $(\forall x)(Fx \supset Gx)$
      2. $+$ $(\exists x)Fx$
      8. ? $(\forall x)Gx$

Let us evaluate this sequence for the class of interpretations incorporating domains consisting of a single object, a. In that domain, the sequence is reducible to the following sequence:

      1. $+$ $Fa \supset Ga$
      2. $+$ Fa
      8. ? Ga

A truth table reveals that no interpretation incorporating a domain consisting of a single object is such that the premises of the sequence are true and its conclusion false

Fa Ga	$Fa \supset Ga$	Fa	Ga
T T	T	T	T
T F	F	T	F
F T	T	F	T
F F	T	F	F

Is the sequence valid? Intuitively, it seems not to be. From "All Fs are Gs" and "Something is an F," it does not appear to follow that "Everything is a G." Let us look at the class of interpretations incorporating domains consisting of two objects, a and b. In such a domain, the sequence is reducible to the sequence

1. $+ (Fa \supset Ga) \wedge (Fb \supset Gb)$
2. $+ Fa \vee Ga$
3. $? \quad Ga \wedge Gb$

Here, the distinction between universally and existentially quantified sentences is signalled by the occurrence of $\wedge$s and $\vee$s, respectively. Suppose we construct a truth table for this sequence that provides a systematic representation of the truth conditions of each sentence for every interpretation that makes use of the domain

Fa	Fb	Ga	Gb	$Fa \supset Ga$	$Fb \supset Gb$	$(Fa \supset Ga) \wedge (Fb \supset Gb)$	$Fa \vee Fb$	$Ga \wedge Gb$
T	T	T	T	T	T	T	T	T
T	T	T	F	T	F	F	T	F
T	T	F	T	F	T	F	T	F
T	T	F	F	F	F	F	T	F
T	F	T	T	T	T	T	T	T
T	F	T	F	T	T	T	T	F
T	F	F	T	F	T	F	T	F
T	F	F	F	F	T	F	T	F
F	T	T	T	T	T	T	T	T
F	T	T	F	T	F	F	T	F
F	T	F	T	T	T	T	T	F
F	T	F	F	T	F	F	T	F
F	F	T	T	T	T	T	F	T
F	F	T	F	T	T	T	F	F
F	F	F	T	T	T	T	F	F
F	F	F	F	T	T	T	F	F

In both the sixth and eleventh rows of the truth table, the premises are true and the conclusion is false. The sequence is invalid: there are interpretations with domains consisting of two individuals in which the premises are true and the conclusion is false. A sequence is invalid if there is *any* interpretation under which its premises are true and its conclusion is false.

In chapter three (§ 3.15) we encountered a less cumbersome method of demonstrating invalidity. We wrote a sequence out horizontally, and then tried to find an assignment of truth values that resulted in the premises being true and the conclusion being false. The technique amounts, in effect, to the construction of a single row of a truth table. Applying this method to the sequence under consideration, we obtain

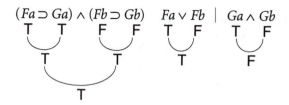

The diagram provides a representation of the sixth row of our original truth table. We can conclude that there is an interpretation, *I*, under which the premises are true and the conclusion false; *I* : {*Fa*=*T*, *Fb*=*F*, *Ga*=*T*, *Gb*=*F*}.

Recall the technique used in the construction of such diagrams. First, we locate assignments of truth values that made the conclusion false. Second, we carry these assignments over to the premises. Third, we attempt to locate assignments of truth values to the remaining elements of the premises that would result in true premises. In the present case, there is more than one assignment of truth values that will make the conclusion false. These assignments correspond to different rows of the original truth table. In some of the rows in which the conclusion is false, one or more of the premises is false. *This means that if, in a given sequence, there is more than one way to make the conclusion false or the premises true, we cannot conclude that the sequence is probably valid solely on the grounds that one of these ways does not result in an interpretation of the sequence that demonstrates its invalidity.* We must first make certain that there is no alternative assignment that will make the premises of the sequence true and its conclusion false.

Suppose we can find no interpretation under which the premises of a sequence are true and its conclusion is false, assuming a domain of two objects. Can we conclude that the sequence is valid? No. There may be an interpretation incorporating a larger domain—or an infinite domain—under which its premises are true and its conclusion false. The truth table technique cannot guarantee success. It can be used to prove invalidity, but not every invalid sequence can be demonstrated invalid using the technique. In general, any invalid sequence whose invalidity would only emerge under interpretations incorporating infinite domains would escape detection.

Despite its limitations, the truth table technique can be extremely useful. Consider the sequence

1. + $(\forall x)Fx \supset (\forall x)Gx$
2. ? $(\forall x)(Fx \supset Gx)$

Is the sequence valid? Maybe not. We might try showing that it is invalid under an interpretation that assumes a domain consisting of a single object, *a*. Assuming that domain, the sequence can be represented as

1. + $Fa \supset Ga$
2. ? $Fa \supset Ga$

Assuming this domain, there is no interpretation under which the conclusion is false and the premise is true: the premise and conclusion are identical, so if the conclusion is false, the premise must be false as well. This is illustrated in the diagram below:

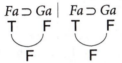

There is no interpretation incorporating a domain consisting of a single object under which the premise is true and the conclusion false. What of interpretations incorporating more populous domains? What of an interpretation incorporating a domain consisting of a *pair* of objects, *a* and *b*? Assuming such a domain, we can represent the original sequence as follows:

1. + $(Fa \wedge Fb) \supset (Ga \wedge Gb)$
2. ? $(Fa \supset Ga) \wedge (Fb \supset Gb)$

We can now see that the premise and conclusion are quite different. The premise says

      If everything is *F* then everything is *G*.

If the "everything" in question is *a* and *b*, then the sentence is reducible to

      If *a* and *b* are *F*, then *a* and *b* are *G*.

The conclusion is

      Everything is such that if it is *F* then it is *G*

that is

If a is F, then a is G and if b is F then b is G.

Assuming this domain, can we find an interpretation under which the premise is true and the conclusion is false?

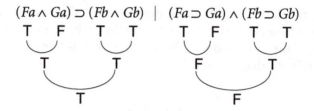

The sequence is shown to be invalid under *I*: {*Fa*=*T*; *Fb*=*T*; *Ga*=*F*; *Gb*=*T*}.

A complication remains. We have seen that, given finite domains, universally quantified sentences are reducible to conjunctions and existentially quantified sentences are reducible to disjunctions. Many sentences in L_p, however, contain *mixed* quantification. How are these to be reduced assuming domains of more than one object? Consider the sentence

$$(\forall x)(\exists y)Mxy$$

The sentence says, in effect, that *everything* bears relation *M* to *something*. Assuming a domain consisting of a single object, *a*, the reduction is simple

$$Maa$$

If the domain includes only a single object, then to say that every object in the domain bears relation *M* to some object is to say that *a* bears *M* to itself. Once again, quantifier differences emerge when we look at reductions for domains consisting of multiple objects. In a two-object domain, the reduction looks like

$$(Maa \lor Mab) \land (Mba \lor Mbb)$$

Given such a domain, the sentence says that *a* bears *M* to *a* or to *b*, and *b* bears *M* to *a* or to *b*. The ∨s carry the sense of the existential quantifier and the ∧s reflect universal quantification. Similarly, assuming our two-object domain, the sentence

$$(\forall x)Fx \supset (\exists y)Gy$$

is reducible to

$$(Fa \wedge Fb) \supset (Ga \vee Gb)$$

and the sentence

$$(\exists x)(Fx \wedge (\forall y)Mxy)$$

reduces to

$$(Fa \wedge (Maa \wedge Mab)) \vee (Fb \wedge (Mba \wedge Mbb))$$

If this seems confusing, think of the original sentence in pidgin $\mathcal{L}_p$: "There is some x such that x is F and, for all y, x bears M to y." If the domain consists of just a and b, then this amounts to: "Either a is F and bears M to both a and b, or b is F and bears M to both a and b."

Although the truth table test for invalidity falls short of perfection, it provides a useful heuristic device for discovering instances of invalidity. In practice, if you suspect that a sequence is invalid, you can often quickly determine whether there is an interpretation under which its conclusion is false and its premises true by assuming finite domains. If such an interpretation emerges, you have proved the sequence invalid. If you can find no such interpretation, however, this may be due either to the fact that the sequence is, after all, valid, or to the fact that its premises are true and its conclusion false only under interpretations involving larger, possibly infinite, domains.

Exercises 5.08

Provide interpretations that demonstrate the invalidity of the sequences below. Use truth tables or charts to support your answer.

1. + (∃x)(Fx ∧ ¬Gx)
   + (∀x)(Hx ⊃ Fx)
   ? (∃x)(Hx ∧ ¬Fx)

2. + (∀x)(Fx ⊃ Gx)
   + (∀x)(¬Fx ⊃ Hx)
   ? (∀x)(¬Gx ⊃ ¬Hx)

3. + ¬(∃x)(Fx ∧ ¬ Gx)
   + ¬(∃x)(Fx ∧ ¬Hx)
   + (∃x)(¬Gx ∧ Jx)
   ? (∃x)((Fx ∧ Gx) ∧ Jx)

4. + (∀x)(Fx ⊃ (∃y)Gxy)
   + (∃x)Fx
   + (∃x)(∃y)Gxy
   ? (∀x)(∃y)Gxy

5. + (∀x)(Fx ⊃ Gx)
   + (∀x)Gx ⊃ (∃x)Hx
   ? (∀x)(Fx ⊃ Hx)

6. + (∃y)(Fy ⊃ (∀x)Gxy)
   + (∀y)(Fy ∧ (∃x)Hyx)
   ? (∃x)(∀y)(Gxy ∧ Hyx)

7. + (∀x)Fx ⊃ (∀x)Gx
   + (∃x)(Fx ∨ Gx)
   ? ¬(∀x)(Fx ⊃ Gx)

8. + (∀x)(Fx ⊃ (Gx ∧ Hx))
   + Fa
   ? (∃x)¬Hx

9. + (∀x)(Fx ⊃ Gxa)
   + (∃x)Gxa ⊃ (∃x)Hxa
   ? (∀x)(∃y)(Fx ⊃ Hxy)

10. + (∃x)(Fx ⊃ (∀y)(Fy ⊃ Jy))
    + (∃x)(Fx ∧ (Gx ∨ Hx))
    ? (∀x)(Fx ⊃ Hx)

11. + (∀x)(∃y)Fxy
    ? (∃y)(∀x)Fxy

12. + (∀x)Fxx
    ? (∀x)(∀y)Fxy

13. + (∃x)(∃y)Fxy
    ? (∃x)Fxx

14. + (∃x)(Fx ∧ Gx)
    + (∀x)(Gx ⊃ Hx)
    ? (∀x)(Gx ∨ Fx)

15. + (∀x)(∃y)Fxy
    + (∃x)Fxx
    ? (∀x)Fxx

5.09 SOUNDNESS AND COMPLETENESS OF L_p

The metalogical notions of soundness and completeness were introduced in
§ 3.17. L_p is *sound* just in case every derivation expressible in L_p is valid:
where $\langle P,c \rangle$ is a sequence, if $P \vdash c$, then $P \vDash c$ (if a set of sentences, P,
deductively yields a sentence, c, then P logically implies c). L_p is *complete* just
in case, if a sequence is valid, a derivation can be given for it in L_p: if $P \vDash c$,
then $P \vdash c$ (if a set of sentences, P, logically implies a sentence, c, then P
deductively yields c).

In our discussion of L_s, no attempt was made to provide exhaustive
proofs of either soundness or completeness. We noted merely what such
proofs involve. The same informal mode of exposition will be used here.

Let us suppose that the soundness of the rules used in L_s derivations has
been established. If L_s is sound, then its rules are *truth preserving*: if we start
with true sentences, applications of those rules to these sentences will yield
only true sentences. Derivations in L_p make use of these same rules, so we
can assume that insofar as an L_p derivation involves only our original set of
rules, L_p must be sound. To extend the proof of soundness, we must show
that the quantifier rules (QT, UI, UG, EI, and EG) and the rule for identity
(ID) are truth preserving: if we apply them to true sentences, they will yield
only true sentences. This amounts to showing that for any sentence ϕ, if ϕ
occurs on a line of a derivation as the result of the application of one of
these rules, then ϕ is logically implied by the sentence or sentences to which
the rule is applied to yield ϕ.

Suppose, for instance, that Γ (*gamma*) is a derivation whose last line is ϕ;
and suppose that Γ yields ϕ by an application of *UI*. Thus, Γ might include
the sentence

$$(\forall x)Fx$$

and ϕ might be the sentence

$$Fa$$

If rule *UI* is truth preserving, there is no interpretation under which the first
sentence is true and the second sentence is false. Given the account of what
it means for a universally quantified sentence like $(\forall x)Fx$ to be true under
an interpretation (see § 4.14), it is plain that if the first sentence is true, the
second must be true as well. This will be so for any pair of sentences
consisting of a universally quantified sentence, $(\forall x)\phi$, and a sentence, $\phi \alpha/\beta$,
derived from that sentence via an application of *UI*.

By extending this line of reasoning to the remaining rules, we produce a proof of the soundness of L_p: ϕ is derivable from Γ only if Γ implies ϕ. A proof for the completeness of L_p is more complicated. We cannot simply extend the truth table technique mentioned in § 3.17 to cover sequences containing quantified sentences. Let us see how a proof might go.

Our aim is to establish that every valid sequence in L_p is derivable using our rules of derivation. Let us say that a set of sentences, Γ, is *consistent with respect to derivability* or, for short, *d-consistent*, just in case the sentence $c \wedge \neg c$ is not derivable from Γ. The notion of *d*-consistency is to be distinguished from consistency *tout court*, our ordinary semantic notion of consistency. If a set of sentences, Γ, is consistent (in this semantic sense), it does not *imply* a contradiction. If Γ is *d*-consistent, a contradiction cannot be *derived from* Γ. A pivotal move in the proof of completeness is a proof that *d*-consistent sets of sentences are consistent in the semantic sense.

We can prove that for any sentence, ϕ, and any set of sentences, Γ, ϕ is derivable from Γ if and only if $\{\Gamma, \neg\phi\}$ is not *d*-consistent. How so? Suppose, first, that ϕ is derivable from Γ, and imagine a derivation of ϕ from Γ in which ϕ appears on the last line. Suppose we supplement Γ with $\neg\phi$. We can now derive $c \wedge \neg c$, so $\{\Gamma, \neg\phi\}$ is not *d*-consistent. In general, once we show that a sequence yields a sentence *and* its contradiction, we can show that it yields any sentence

$$\vdots$$

i. ϕ
j. $\neg\phi$
k. $\phi \vee (c \wedge \neg c)$ i, vI
l. $(c \wedge \neg c)$ j, k, $\wedge E$

Now suppose that $\{\Gamma, \neg\phi\}$ is not *d*-consistent. In that case, $c \wedge \neg c$ is derivable from $\{\Gamma, \neg\phi\}$. If that is so, we can derive the sentence

$$\neg\phi \supset (c \wedge \neg c)$$

from Γ. (In general, for any set of sentences, Γ, and any sentences ϕ and c, if $\{\Gamma, \phi\} \vdash c$, then $\Gamma \vdash \phi \supset c$.) Having derived this sentence, we can derive ϕ from Γ by deriving $\neg(c \wedge \neg c)$ by means of *IP*, and then applying *MP*

$$\vdots$$

i. $\neg\phi \supset (c \wedge \neg c)$
j. $\lceil c \wedge \neg c$
k. $\lvert\ ? \times$

$$
\begin{array}{lll}
\text{l.} & \quad c & j, \wedge E \\
\text{m.} & \quad \neg c & j, \wedge E \\
\text{n.} & \quad c \wedge \neg c & \text{l, m, } \wedge I \\
\text{o.} & \neg(c \wedge \neg c) & \text{j–n, } IP \\
\text{l.} & \phi & \text{i, o, } MT
\end{array}
$$

Given the notion of d-consistency, we can define a related notion of *maximal d-consistency* as follows: a set of sentences, Γ, is maximally d-consistent if and only if (i) Γ is d-consistent, and (ii) Γ is not a proper subset of any d-consistent set of sentences. If Γ is maximally d-consistent, then, if ϕ is a sentence that is not in Γ, $\{\Gamma, \phi\}$ is not d-consistent.

Suppose now that Γ is maximally d-consistent and that ϕ is a sentence of $\mathcal{L}_p$. It follows that ϕ is a member of Γ if and only if $\neg\phi$ is *not* a member of Γ. It follows, as well, that ϕ is a member of Γ if and only if ϕ is derivable from Γ. (The derivation in this case will be trivial: if ϕ is *in* Γ, then ϕ can be derived from Γ simply by deriving $\phi \vee \phi$ from ϕ and then applying *Taut*.) Given these results, it can be shown that

1. $\phi \wedge \psi$ is in Γ if and only if ϕ is in Γ and ψ is in Γ;
2. $\phi \vee \psi$ is in Γ if and only if ϕ is in Γ, or ψ is in Γ, or both;
3. $\phi \supset \psi$ is in Γ if and only if ϕ is not in Γ, or ψ is in Γ, or both;
4. $\phi \equiv \psi$ is in Γ if and only if ϕ is in Γ and ψ is in Γ, or neither ϕ nor ψ is in Γ.

Let us call Γ ω-complete (*omega*-complete) just in case, for every expression ϕ and every variable, α, if $(\exists\alpha)\phi$ is in Γ, then $\phi\alpha/\beta$ is in Γ (where $\phi\alpha/\beta$ is the sentence that results from dropping the quantifier in $(\exists\alpha)\phi$ and replacing each instance of the variable α bound by that quantifier with some individual constant, β). If Γ is ω-complete, then, Γ contains an existentially quantified sentence if and only if it contains, as well, a sentence from which that existentially quantified sentence can be derived by an application of rule *EG*. If Γ is both maximally d-consistent and ω-complete, then the following will hold:

5. $(\forall\alpha)\phi$ is in Γ if and only if for every individual constant, β, $\phi\alpha/\beta$ is in Γ;
6. $(\exists\alpha)\phi$ is in Γ if and only if there is an individual constant, β, such that $\phi\alpha/\beta$ is in Γ.

Given all this, we can advance a proof for the completeness of $\mathcal{L}_p$. To do so, we first show that the following is true:

Every d-consistent set of sentences is consistent.

Once this link between the semantic notion of consistency and the syntactic notion of d-consistency is forged, completeness can be proved as follows. Suppose that Γ is a set of sentences and ϕ is logically implied by Γ (i.e., $\Gamma \vDash \phi$). Then $\{\Gamma, \neg\phi\}$ is not consistent, and so not d-consistent. If $\{\Gamma, \neg\phi\}$ is not d-consistent, the ϕ is derivable from Γ (i.e., $\Gamma \vdash \phi$). Thus $\mathcal{L}_p$ is complete.

The foregoing constitutes only an outline of a full proof of completeness. For a more detailed account, see B. Mates, *Elementary Logic* 2d ed. (New York: Oxford University Press, 1972), pp. 142–47. The discussion here is intended to provide a hint of a deeper and richer territory lying beneath the formal languages that have occupied us. Those languages represent the tip of an iceberg. If this book has been successful, it has provided you with a feel for the formal languages and a sense of what lies beneath the surface.

Selected Solutions to Exercises

CHAPTER TWO EXERCISES

Exercises 2.00 (p. 11)

  2. S 4. F

Exercises 2.01 (p. 20)

  2. $H \wedge G$ 4. $\neg(F \wedge \neg H)$
  6. $\neg(G \wedge H)$ 8. $\neg H \wedge \neg G$
 10. Ambiguous: $\neg((H \wedge G) \wedge \neg F)$ or $\neg(H \wedge G) \wedge \neg F$

Exercises 2.02 (p. 25)

  2. $H \vee \neg G$ 4. $\neg(F \vee \neg H)$
  6. $(G \vee H) \wedge \neg(G \wedge H)$ 8. $\neg H \vee \neg G$
 10. $E \vee (G \vee H)$

Exercises 2.03 (p. 36)

  2. $E \supset H$ 4. $(G \vee H) \supset F$
  6. $E \supset (G \supset F)$ 8. $E \supset (H \vee G)$
 10. $F \supset (E \vee (G \vee H))$

Exercises 2.04 (p. 39)

  2. $H \supset E$ 4. $H \equiv (E \vee F)$
  6. $E \equiv (G \wedge H)$ 8. $(H \vee G) \equiv E$
 10. $F \supset (E \equiv \neg G)$

Exercises 2.05 (p. 46)

2.

P Q	$\neg P$	$\neg Q$	$\neg P \wedge \neg Q$
T T	F	F	**F**
T F	F	T	**F**
F T	T	F	**F**
F F	T	T	**T**

4.

P Q	$\neg P$	$\neg Q$	$\neg P \vee \neg Q$	$\neg(\neg P \vee \neg Q)$
T T	F	F	F	**T**
T F	F	T	T	**F**
F T	T	F	T	**F**
F F	T	T	T	**F**

6.

P Q	$\neg Q$	$P \supset \neg Q$
T T	F	**F**
T F	T	**T**
F T	F	**T**
F F	T	**T**

8.

P Q R	$\neg R$	$Q \wedge \neg R$	$\neg(Q \wedge \neg R)$	$P \supset \neg(Q \wedge \neg R)$
T T T	F	F	T	**T**
T T F	T	T	F	**F**
T F T	F	F	T	**T**
T F F	T	F	T	**T**
F T T	F	F	T	**T**
F T F	T	T	F	**T**
F F T	F	F	T	**T**
F F F	T	F	T	**T**

10.

P Q R	¬P	¬Q	¬R	¬Q∧¬R	¬P∨(¬Q∧¬R)	¬(¬P∨(¬Q∧¬R))
T T T	F	F	F	F	F	**T**
T T F	F	F	T	F	F	**T**
T F T	F	T	F	F	F	**T**
T F F	F	T	T	T	T	**F**
F T T	T	F	F	F	T	**F**
F T F	T	F	T	F	T	**F**
F F T	T	T	F	F	T	**F**
F F F	T	T	T	T	T	**F**

Exercises 2.06 (p. 49)

2.

P Q	P⊃Q	¬Q	P∧¬Q	¬(P∧¬Q)
T T	**T**	F	F	**T**
T F	**F**	T	T	**F**
F T	**T**	F	F	**T**
F F	**T**	T	F	**T**

4.

P Q	P∨Q	P∣Q	(P∣Q)∣(P∣Q)
T T	**T**	F	**T**
T F	**T**	F	**T**
F T	**T**	F	**T**
F F	**F**	T	**F**

6.

P Q	P∨Q	¬P	¬Q	¬P∧¬Q	¬(¬P∧¬Q)
T T	**T**	F	F	F	**T**
T F	**T**	F	T	F	**T**
F T	**T**	T	F	F	**T**
F F	**F**	T	T	T	**F**

8.

P Q	P⊃Q	¬P	¬P∨Q
T T	**T**	F	**T**
T F	**F**	F	**F**
F T	**T**	T	**T**
F F	**T**	T	**T**

10.

P Q	P∣Q	¬P	¬P⊃Q	¬(¬P⊃Q)
T T	**F**	F	T	**F**
T F	**F**	F	T	**F**
F T	**F**	T	T	**F**
F F	**T**	T	F	**T**

Exercises 2.07 (p. 54)

2. $J \wedge C$

4. $\neg(F \wedge G)$

6. $(F \equiv J) \wedge \neg G$

8. $(J \vee \neg C) \supset G$

10. $(G \wedge \neg F) \wedge C$

Exercises 2.08 (p. 57)

2. $\neg(I \vee G)$

4. $\neg(I \vee G) \supset F$

6. $\neg(I \wedge G)$

8. $\neg C \supset (I \vee G)$

10. $(\neg C \wedge J) \supset (I \vee G)$

Exercises 2.09 (p. 61)

2. $\neg G \supset \neg J$

4. $J \supset G$

6. $J \supset G$

8. $F \equiv (C \wedge \neg J)$

10. $J \supset (F \equiv \neg G)$

Exercises 2.10 (pp. 65–66)

2. $I \vee G$

4. $\neg F \vee C$

6. $I \supset (J \vee G)$

8. $(I \wedge \neg G) \wedge J$

10. $(C \supset F) \wedge J$

Exercises 2.11 (p. 69)

2.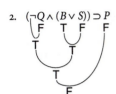

4.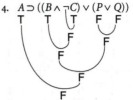

6. ((A ∧ B) ∧ P) ≡ (B ⊃ (C ∨ S))

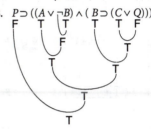

8. P ⊃ ((A ∨ ¬B) ∧ (B ⊃ (C ∨ Q)))

10. ¬((B ∨ (P ∧ ¬Q)) ⊃ ((A ∧ ¬B) ∨ (¬P ⊃ (Q ∨ R))))

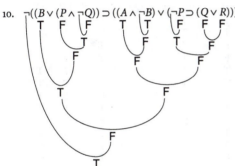

Exercises 2.12 (p. 74)

2. logically true (tautologous)   4. contradiction
6. logically true (tautologous)   8. contradiction
10. logically true (tautologous)

CHAPTER THREE EXERCISES

Exercises 3.00 (p. 86)

2. "One plus one" is not identical with "two."   4. "Sincerity" involves "sin."
6. "One" is not identical with "one."   8. Sentences 8 and 9 cannot both be true.
10. I love the sound of "a cellar door."

Exercises 3.01 (p. 91)

2. invalid (sixth row)   4. valid
6. invalid (third row)   8. invalid (fifth row)
10. invalid (second row)

P Q R	② P⊃Q	P⊃R	Q⊃R	⑧ P⊃Q	R⊃Q	R	P
T T T	T	T	T	T	T	T	T
T T F	T	F	F	T	T	F	T
T F T	F	T	T	F	F	T	T
T F F	F	F	T	F	T	F	T
F T T	T	T	T	T	T	T	F
F T F	T	T	F	T	T	F	F
F F T	T	T	T	T	F	T	F
F F F	T	T	T	T	T	F	F

P Q	⑥ Q⊃P	P⊃(Q⊃P)	P	⑩ P∨Q	P	¬Q
T T	T	T	T	T	T	F
T F	T	T	T	T	T	T
F T	F	T	F	T	F	F
F F	T	T	F	F	F	T

④

P Q R S	¬P	¬Q	¬R	¬S	Q∧¬R	¬(Q∧¬R)	P∨¬Q	¬(Q∧¬R)⊃¬S	¬P	¬Q
T T T T	F	F	F	F	F	T	T	F	F	F
T T T F	F	F	F	T	F	T	T	T	F	F
T T F T	F	F	T	F	T	F	T	T	F	F
T T F F	F	F	T	T	T	F	T	T	F	F
T F T T	F	T	F	F	F	T	T	F	F	T
T F T F	F	T	F	T	F	T	T	T	F	T
T F F T	F	T	T	F	F	T	T	F	F	T
T F F F	F	T	T	T	F	T	T	T	F	T
F T T T	T	F	F	F	F	T	F	F	T	F
F T T F	T	F	F	T	F	T	F	T	T	F
F T F T	T	F	T	F	T	F	F	T	T	F
F T F F	T	F	T	T	T	F	F	T	T	F
F F T T	T	T	F	F	F	T	T	F	T	T
F F T F	T	T	F	T	F	T	T	T	T	T
F F F T	T	T	T	F	F	T	T	F	T	T
F F F F	T	T	T	T	F	T	T	T	T	T

Exercises 3.02 (p. 102)

2. 1.+ ¬P⊃¬Q
   2.+ ¬P
   3. ? ¬Q
   4.   ¬Q   1, 2, MP

4. 1.+ P⊃Q
   2.+ ¬S
   3.+ ¬(Q⊃R)⊃S
   4. ? P⊃R
   5.   Q⊃R   2, 3, MT
   6.   P⊃R   1, 5, HS

6. 1.+ P⊃(Q⊃R)
   2.+ P
   3.+ Q
   4. ? R
   5.   Q⊃R   1, 2, MP
   6.   R   3, 5, MP

8. 1.+ P⊃Q
   2.+ Q⊃R
   3.+ P
   4. ? R
   5.   P⊃R   1, 2, HS
   6.   R   3, 5, MP

10. 1.+ $\neg(P \supset R) \supset \neg Q$
     2.+ P
     3.+ Q
     4. ? R
     5. $P \supset R$ 1, 3, MT
     6. R 2, 5, MP

Exercises 3.03 (pp. 104–105)

2. 1.+ $P \wedge (\neg Q \wedge \neg R)$ 4. 1.+ $\neg P$
    2. ? $\neg R$ 2.+ Q
    3. $\neg Q \wedge \neg R$ 1, $\wedge E$ 3.+ $(\neg P \wedge Q) \supset R$
    4. $\neg R$ 3, $\wedge E$ 4. ? R
                                                        5. $\neg P \wedge Q$ 1, 2, $\wedge I$
                                                        6. R 3, 5, MP

6. 1.+ $P \supset (Q \wedge \neg R)$ 8. 1.+ $(P \wedge Q) \supset (R \wedge S)$
    2.+ P 2.+ Q
    3. ? $\neg R$ 3.+ P
    4. $Q \wedge \neg R$ 1, 2, MP 4. ? R
    5. $\neg R$ 4, $\wedge E$ 5. $P \wedge Q$ 2, 3, $\wedge I$
                                                         6. $R \wedge S$ 1, 5, MP
                                                         7. R 6, $\wedge E$

10. 1.+ $S \wedge ((P \equiv Q) \supset R)$
     2.+ $P \equiv Q$
     3. ? R
     4. $(P \equiv Q) \supset R$ 1, $\wedge E$
     5. R 2, 4, MP

Exercises 3.04 (pp. 108–109)

2. 1.+ $(P \vee Q) \supset (R \wedge S)$ 4. 1.+ $P \supset (Q \vee R)$
    2.+ P 2.+ $\neg(Q \vee R) \vee S$
    3. ? S 3.+ $\neg S$
    4. $P \vee Q$ 2, $\vee I$ 4. ? $\neg P$
    5. $R \wedge S$ 1, 4, MP 5. $\neg(Q \vee R)$ 2, 3, $\vee E$
    6. S 5, $\wedge E$ 6. $\neg P$ 1, 5, MT

6. 1.+ $P \supset \neg(Q \wedge R)$ 8. 1.+ $P \supset (Q \vee \neg S)$
    2.+ $(Q \wedge R) \vee S$ 2.+ $P \wedge S$
    3.+ $\neg S$ 3. ? Q
    4. ? $\neg P$ 4. P 2, $\wedge E$
    5. $Q \wedge R$ 2, 3, $\vee E$ 5. $Q \vee \neg S$ 1, 4, MP
    6. $\neg P$ 1, 5, MT 6. S 2, $\wedge E$
                                                        7. Q 5, 6, $\vee E$

10. 1.+ $P \supset (Q \wedge R)$
   2.+ $S \vee \neg T$
   3.+ $S \supset P$
   4.+ T
   5. ? Q
   6. S 2, 4, $\vee E$
   7. P 3, 6, MP
   8. $Q \wedge R$ 1, 7, MP
   9. Q 8, $\wedge E$

Exercises 3.05 (pp. 112–113)

2. 1.+ $P \supset (S \supset (Q \wedge R))$
   2.+ $(Q \wedge R) \supset \neg P$
   3.+ $T \supset S$
   4. ? $P \supset \neg T$
   5. $\lceil P$
   6. $| ? \neg T$
   7. $| S \supset (Q \wedge R)$ 1, 5, MP
   8. $| \neg(Q \wedge R)$ 2, 5, MP
   9. $| \neg S$ 7, 8, MT
  10. $\lfloor \neg T$ 3, 9, MT
  11. $P \supset \neg T$ 5–10, CP

4. 1.+ $(P \wedge Q) \supset R$
   2.+ P
   3. ? $Q \supset R$
   4. $\lceil Q$
   5. $| ? R$
   6. $| P \wedge Q$ 2, 4, $\wedge I$
   7. $\lfloor R$ 1, 6, MP
   8. $Q \supset R$ 4–7, CP

6. 1.+ $P \supset (Q \vee R)$
   2.+ $P \supset \neg Q$
   3. ? $P \supset R$
   4. $\lceil P$
   5. $| ? R$
   6. $| Q \vee R$ 1, 4, MP
   7. $| \neg Q$ 2, 4, MP
   8. $\lfloor R$ 6, 7, $\vee E$
   9. $P \supset R$ 4–8, CP

8. 1.+ $P \supset S$
   2.+ $R \supset S$
   3. ? $P \supset (R \supset S)$
   4. $\lceil P$
   5. $| ? R \supset S$
   6. $| \lceil R$
   7. $| | ? S$
   8. $| \lfloor S$ 1, 4, MP
   9. $\lfloor R \supset S$ 6–8, CP
  10. $P \supset (R \supset S)$ 4–9, CP

10. 1.+ $Q \supset (T \vee S)$
   2.+ $\neg R \wedge \neg T$
   3.+ P
   4. ? $P \wedge (Q \supset S)$
   5. $\lceil Q$
   6. $| ? S$
   7. $| T \vee S$ 1, 5, MP
   8. $| \neg T$ 2, $\wedge E$
   9. $\lfloor S$ 7, 8, $\vee E$
  10. $Q \supset S$ 5–9, CP
  11. $P \wedge (Q \supset S)$ 3, 10, $\wedge I$

Exercises 3.06 (p. 116)

2. 1.+ $P \lor Q$
    2.+ $P \supset (R \land S)$
    3.+ $(R \lor S) \supset Q$
    4. ? Q
    5. $\neg Q$
    6.   ? X
    7. P 1, 5, $\lor E$
    8. $R \land S$ 2, 7, MP
    9. $\neg(R \land S)$ 3, 5, MT
    10. $(R \land S) \land \neg(R \land S)$ 8, 9, $\land I$
    11. Q 5–10, IP

4. 1.+ $\neg R \supset \neg(\neg P \lor Q)$
    2.+ $\neg R$
    3. ? P
    4. $\neg P$
    5.   ? X
    6. $\neg(\neg P \lor Q)$ 1, 2, MP
    7. $\neg P \lor Q$ 4, $\lor I$
    8. $(\neg P \lor Q) \land \neg(\neg P \lor Q)$ 6, 7, $\land I$
    9. P 4–8, IP

6. 1.+ $P \supset Q$
    2.+ $S \supset T$
    3. ? $(P \lor S) \supset \neg(\neg Q \land \neg T)$
    4. $P \lor S$
    5.   ? $\neg(\neg Q \land \neg T)$
    6. $\neg Q \land \neg T$
    7.   ? X
    8. $\neg Q$ 6, $\land E$
    9. $\neg P$ 1, 8, MT
    10. S 4, 9, $\lor E$
    11. $\neg T$ 6, $\land E$
    12. $\neg S$ 2, 11, MT
    13. $S \land \neg S$ 10, 12, $\land I$
    14. $\neg(\neg Q \land \neg T)$ 6–13, IP
    15. $(P \lor S) \supset \neg(\neg Q \land \neg T)$ 4–14, CP

8. 1.+ $(P \lor Q) \supset (R \supset S)$
    2.+ $\neg P \supset T$
    3.+ $R \land \neg S$
    4. ? T
    5. $\neg T$
    6.   ? X
    7. P 2, 5, MP
    8. $P \lor Q$ 7, $\lor I$
    9. $R \supset S$ 1, 8, MP
    10. R 3, $\land E$
    11. S 9, 10, MP
    12. $\neg S$ 3, $\land E$
    13. $S \land \neg S$ 11, 12, $\land I$
    14. T 5–13, IP

10. 1.+ $P \supset (\neg Q \land R)$
    2.+ $S \lor \neg T$
    3.+ $P \lor T$
    4. ? $Q \supset S$
    5. Q
    6.   ?S
    7. $\neg S$
    8.   ? X
    9. $\neg T$ 2, 7, $\lor E$
    10. P 3, 9, $\lor E$
    11. $\neg Q \land R$ 1, 10, MP
    12. $\neg Q$ 11, $\land E$
    13. $Q \land \neg Q$ 5, 12, $\land I$
    14. S 7–13, IP
    15. $Q \supset S$ 5–14, CP

Exercises 3.07 (p. 123)

2. 1.+ $(P \supset Q) \vee (R \vee S)$
    2. ? $(S \vee R) \vee P \supset Q$
    3. $(R \vee S) \vee (P \supset Q)$ 1, *Com*
    4. $(S \vee R) \vee (P \supset Q)$ 3, *Com*

4. 1.+ $(P \vee Q) \wedge (R \wedge S)$
    2. ? $(R \wedge (P \vee Q)) \wedge S$
    3. $((P \vee Q) \wedge R) \wedge S$ 1, *Assoc*
    4. $(R \wedge (P \vee Q)) \wedge S$ 3, *Com*

6. 1.+ $P \supset ((R \vee Q) \supset S)$
    2.+ $(T \vee S) \supset W$
    3. ? $P \supset (Q \supset W)$
    4.   ⌐ P
    5.   │ ? $Q \supset W$
    6.   │ ⌐ Q
    7.   │ │ ? W
    8.   │ │ $(R \vee Q) \supset S$ 1, 4, *MP*
    9.   │ │ $Q \vee R$ 6, $\vee I$
    10. │ │ $R \vee Q$ 9, *Com*
    11. │ │ S 8, 10, *MP*
    12. │ │ $S \vee T$ 11, $\vee I$
    13. │ │ $T \vee S$ 12, *Com*
    14. │ │ W 2, 13, *MP*
    15. │ ⌐ $Q \supset W$ 6–14, *CP*
    16. $P \supset (Q \supset W)$ 4–15, *CP*

8. 1.+ $(P \vee (Q \vee R)) \supset T$
    2.+ $(S \vee \neg T) \supset R$
    3. ? T
    4.   ⌐ $\neg T$
    5.   │ ? X
    6.   │ $\neg T \vee S$ 4, $\vee I$
    7.   │ $S \vee \neg T$ 6, *Com*
    8.   │ R 2, 7, *MP*
    9.   │ $R \vee Q$ 8, $\vee I$
    10. │ $Q \vee R$ 9, *Com*
    11. │ $(Q \vee R) \vee P$ 10, $\vee I$
    12. │ $P \vee (Q \vee R)$ 11, *Com*
    13. │ T 1, 12, *MP*
    14. ⌐ $T \wedge \neg T$ 4, 13, $\wedge I$
    15. T 4–14, *IP*

10. 1.+ $P \supset ((Q \wedge R) \vee S)$
    2.+ $(R \wedge Q) \supset \neg P$
    3.+ $T \supset \neg S$
    4. ? $P \supset \neg T$
    5.   ⌐ P
    6.   │ ? $\neg T$
    7.   │ $(Q \wedge R) \vee S$ 1, 5, *MP*
    8.   │ ⌐ T
    9.   │ │ ? X
    10. │ │ $\neg S$ 3, 9, *MT*
    11. │ │ $Q \wedge R$ 7, 10, $\vee E$
    12. │ │ $\neg (R \wedge Q)$ 2, 5, *MT*
    13. │ │ $\neg (Q \wedge R)$ 12, *Com*
    14. │ ⌐ $(Q \wedge R) \wedge \neg (Q \wedge R)$ 11, 13, $\wedge I$
    15. │ ⌐ $\neg T$ 8–14, *IP*
    16. $P \supset \neg T$ 5–15, *CP*

Exercises 3.08 (p. 126)

2. 1.+ $\neg (\neg P \vee (\neg Q \wedge \neg R))$
    2. ? $P \wedge (Q \vee R)$
    3. $P \wedge \neg (\neg Q \wedge \neg R)$ 1, *DeM*
    4. $P \wedge (Q \vee R)$ 3, *DeM*

4. 1.+ $P \supset \neg (Q \wedge (R \vee \neg S))$
    2. ? $P \supset (\neg Q \vee (\neg R \wedge S))$
    3. $P \supset (\neg Q \vee \neg (R \vee \neg S))$ 1, *DeM*
    4. $P \supset (\neg Q \vee (\neg R \wedge S))$ 3, *DeM*

6. 1.+ $Q \supset S$
    2.+ $S \supset P$
    3.? $P \vee \neg Q$
    4. $\neg(P \vee \neg Q)$
    5. $? X$
    6. $\neg P \wedge Q$ 4, *DeM*
    7. $\neg P$ 6, $\wedge E$
    8. $\neg S$ 2, 7, *MT*
    9. $\neg Q$ 1, 8, *MT*
    10. Q 6, $\wedge E$
    11. $Q \wedge \neg Q$ 9, 10, $\wedge I$
    12. $P \vee \neg Q$ 4–11, *IP*

8. 1.+ $P \supset (Q \vee S)$
    2.+ $\neg S$
    3.? $\neg P \vee Q$
    4. $\neg(\neg P \vee Q)$
    5. $? X$
    6. $P \wedge \neg Q$ 4, *DeM*
    7. P 6, $\wedge E$
    8. $Q \vee S$ 1, 7, *MP*
    9. $\neg Q$ 6, $\wedge E$
    10. S 8, 9, $\vee E$
    11. $S \wedge \neg S$ 2, 10, $\wedge I$
    12. $\neg P \vee Q$ 4–11, *IP*

10. 1.+ $P \vee (Q \vee R)$
    2.+ $Q \supset (R \wedge S)$
    3.? $P \vee R$
    4. $\neg(P \vee R)$
    5. $? X$
    6. $\neg P \wedge \neg R$ 4, *DeM*
    7. $\neg P$ 6, $\wedge E$
    8. $Q \vee R$ 1, 7, $\vee E$
    9. $\neg R$ 6, $\wedge E$
    10. Q 8, 9, $\vee E$
    11. $R \wedge S$ 2, 10, *MP*
    12. R 11, $\wedge E$
    13. $R \wedge \neg R$ 9, 12, $\wedge I$
    14. $P \vee R$ 4–13, *IP*

Exercise 3.09 (pp. 129–30)

2. 1.+ $\neg(P \vee Q) \supset R$
    2.? $\neg P \supset (\neg Q \supset R)$
    3. $(\neg P \wedge \neg Q) \supset R$ 1, *DeM*
    4. $\neg P \supset (\neg Q \supset R)$ 3, *Exp*

4. 1.+ $P \vee (Q \wedge \neg R)$
    2.? $(P \vee Q) \wedge \neg(\neg P \wedge R)$
    3. $(P \vee Q) \wedge (P \vee \neg R)$ 1, *Dist*
    4. $(P \vee Q) \wedge \neg(\neg P \wedge R)$ 3, *DeM*

6. 1.+ $P \supset (Q \wedge R)$
    2.+ $Q \supset (R \supset S)$
    3.? $P \supset S$
    4. $(R \wedge Q) \supset S$ 2, *Exp*
    5. $(Q \wedge R) \supset S$ 4, *Com*
    6. $P \supset S$ 1, 5, *HS*

8. 1.+ $(P \wedge R) \supset Q$
    2.+ $P \supset R$
    3.? $P \supset Q$
    4. $(R \wedge P) \supset Q$ 1, *Com*
    5. $R \supset (P \supset Q)$ 4, *Exp*
    6. $P \supset (P \supset Q)$ 2, 5, *HS*
    7. $(P \wedge P) \supset Q$ 6, *Exp*
    8. $P \supset Q$ 7, *Taut*

10. 1.+ $(S \lor T) \supset (\neg P \lor \neg R)$
     2.+ $S \lor (Q \land T)$
     3. ? $P \supset \neg R$
     4. ⌈ P
     5. | ? $\neg R$
     6. | $(S \lor Q) \land (S \lor T)$ 2, Dist
     7. | $S \lor T$ 6, $\land E$
     8. | $\neg P \lor \neg R$ 1, 7, MP
     9. ⌊ $\neg R$ 4, 8, $\lor E$
     10. $P \supset \neg R$ 4–9, CP

Exercise 3.10 (pp. 132–33)

2. 1.+ $\neg P \supset (Q \land R)$
    2. ? $\neg (Q \land R) \supset P$
    3. $\neg (Q \land R) \supset P$ 1, Contra

4. 1.+ $\neg P \supset (\neg Q \supset R)$
    2. ? $(P \lor Q) \lor R$
    3. $(\neg P \land \neg Q) \supset R$ 1, Exp
    4. $\neg (\neg P \land \neg Q) \lor R$ 3, Cond
    5. $(P \lor Q) \lor R$ 4, DeM

6. 1.+ $(\neg S \lor R) \supset (T \supset P)$
    2.+ $S \supset R$
    3. ? $\neg T \lor P$
    4. $\neg S \lor R$ 2, Cond
    5. $T \supset P$ 1, 4, MP
    6. $\neg T \lor P$ 5, Cond

8. 1.+ $((P \supset Q) \supset R) \supset S$
    2.+ R
    3. ? S
    4. ⌈ $\neg S$
    5. | ? X
    6. | $\neg ((P \supset Q) \supset R)$ 1, 4, MT
    7. | $\neg (\neg (P \supset Q) \lor R)$ 6, Cond
    8. | $(P \supset Q) \land \neg R$ 7, DeM
    9. | $\neg R$ 8, $\land E$
    10. ⌊ $R \land \neg R$ 2, 9, $\land I$
    11. S 4–10, IP

10. 1.+ $\neg P \lor R$
     2.+ $(P \land \neg R) \lor S$
     3.+ $(R \land S) \supset Q$
     4. ? $P \supset Q$
     5. ⌈ P
     6. | ? Q
     7. | $\neg (P \land \neg R) \supset S$ 2, Cond
     8. | $(\neg P \lor R) \supset S$ 7, DeM
     9. | S 1, 8, MP
     10. | R 1, 5, $\lor E$
     11. | $R \land S$ 9, 10, $\land I$
     12. ⌊ Q 3, 11, MP
     13. $P \supset Q$ 5–12, CP

Exercises 3.11 (pp. 134–35)

2. 1.+ $P \equiv Q$
    2. ? $(P \lor \neg Q) \land (\neg P \lor Q)$
    3. $(P \supset Q) \land (Q \supset P)$ 1, Bicond
    4. $(\neg P \lor Q) \land (Q \supset P)$ 3, Cond
    5. $(\neg P \lor Q) \land (\neg Q \lor P)$ 4, Cond
    6. $(\neg Q \lor P) \land (\neg P \lor Q)$ 5, Com
    7. $(P \lor \neg Q) \land (\neg P \lor Q)$ 6, Com

4. 1.+ $(P \supset Q) \land \neg (\neg P \land Q)$
    2. ? $P \equiv Q$
    3. $(P \supset Q) \land (P \lor \neg Q)$ 1, DeM
    4. $(P \supset Q) \land (\neg Q \lor P)$ 3, Com
    5. $(P \supset Q) \land (Q \supset P)$ 4, Cond
    6. $P \equiv Q$ 5, Bicond

6.
1.+	$(P \lor Q) \supset (R \equiv \neg S)$	
2.+	$(S \lor T) \supset (P \land R)$	
3. ?	$\neg S$	
4.	S	
5.	$? X$	
6.	$S \lor T$	2, $\lor I$
7.	$P \land R$	2, 6, MP
8.	P	7, $\land E$
9.	$P \lor Q$	8, $\lor I$
10.	$R \equiv \neg S$	1, 9, MP
11.	$(R \supset \neg S) \land (\neg S \supset R)$	10, $Bicond$
12.	$R \supset \neg S$	11, $\land E$
13.	$\neg R$	4, 12, MT
14.	R	7, $\land E$
15.	$R \land \neg R$	13, 14, $\land I$
16.	$\neg S$	4–15, IP

8.
1.+	$P \supset (Q \equiv R)$	
2.+	$\neg S \supset (P \lor R)$	
3.+	$P \equiv Q$	
4. ?	$S \lor R$	
5.	$\neg (S \lor R)$	
6.	$? X$	
7.	$\neg S \land \neg R$	5, DeM
8.	$\neg S$	7, $\land E$
9.	$P \lor R$	2, 8, MP
10.	$\neg R$	7, $\land E$
11.	P	9, 10, $\lor E$
12.	$Q \equiv R$	1, 11, MP
13.	$(Q \supset R) \land (R \supset Q)$	12, $Bicond$
14.	$(P \supset Q) \land (Q \supset P)$	3, $Bicond$
15.	$P \supset Q$	14, $\land E$
16.	$Q \supset R$	13, $\land E$
17.	$P \supset R$	15, 16, HS
18.	R	11, 17, MP
19.	$R \land \neg R$	10, 18, $\land I$
20.	$S \lor R$	5–19, IP

10.
1.+	$S \equiv T$	
2.+	$S \supset (P \lor Q)$	
3. ?	$\neg Q \supset (T \supset P)$	
4.	$\neg Q$	
5.	$? T \supset P$	
6.	T	
7.	$? P$	
8.	$(S \supset T) \land (T \supset S)$	1, $Bicond$
9.	$T \supset S$	8, $\land E$
10.	S	6, 9, MP
11.	$P \lor Q$	2, 10, MP
12.	P	4, 11, $\lor E$
13.	$T \supset P$	6–12, CP
14.	$\neg Q \supset (T \supset P)$	4–13, CP

Exercises 3.12 (p. 140)

2.
1.+	$S \supset P$	
2.+	$Q \supset P$	
3.+	$\neg Q \supset S$	
4. ?	P	
5.	$Q \lor S$	3, $Cond.$
6.	$P \lor P$	1, 2, 5, CD
7.	P	6, $Taut$

4.
1.+	$P \supset (R \land T)$	
2.+	$Q \supset (S \land T)$	
3. ?	$(P \lor Q) \supset (R \lor S)$	
4.	$P \lor Q$	
5.	$? R \lor S$	
6.	$(R \land T) \lor (S \land T)$	1, 2, 4, CD
7.	$(T \land R) \lor (S \land T)$	6, Com
8.	$(T \land R) \lor (T \land S)$	7, Com
9.	$T \land (R \lor S)$	8, $Dist$
10.	$R \lor S$	9, $\land E$
11.	$(P \lor Q) \supset (R \lor S)$	4–10, CP

6. 1.+ $P \lor R$
    2.+ $P \supset (Q \land \neg S)$
    3.+ $(\neg R \lor T) \land \neg S$
    4. ? $Q \lor T$
    5. $\neg P \lor (Q \land \neg S)$ 2, *Cond.*
    6. $(\neg P \lor Q) \land (\neg P \lor \neg S)$ 5, *Dist*
    7. $\neg P \lor Q$ 6, $\land E$
    8. $P \supset Q$ 7, *Cond*
    9. $\neg R \lor T$ 3, $\land E$
    10. $R \supset T$ 9, *Cond*
    11. $Q \lor T$ 1, 8, 10, *CD*

8. 1.+ $\neg P$
    2.+ $\neg Q$
    3. ? $(P \lor Q) \supset (R \lor S)$
    4.   ⎡ $P \lor Q$
    5. | ? $R \lor S$
    6. | $\neg P \lor R$ 1, $\lor I$
    7. | $P \supset R$ 6, *Cond*
    8. | $\neg Q \lor S$ 2, $\lor I$
    9. | $Q \supset S$ 8, *Cond*
    10. ⎣ $R \lor S$ 4, 7, 9, *CD*
    11. $(P \lor Q) \supset (R \lor S)$ 4–10, *CP*

10. 1.+ $P \supset (Q \lor R)$
    2.+ $S \supset (R \lor T)$
    3.+ $\neg R$
    4. ? $(P \lor S) \supset (Q \lor T)$
    5.   ⎡ $P \lor S$
    6. | ? $Q \lor T$
    7. | $(Q \lor R) \lor (R \lor T)$ 1, 2, 5, *CD*
    8. | $Q \lor (R \lor (R \lor T))$ 7, *Assoc*
    9. | $Q \lor ((R \lor R) \lor T)$ 8, *Assoc*
    10. | $Q \lor (R \lor T)$ 9, *Taut*
    11. | $Q \lor (T \lor R)$ 10, *Com*
    12. | $(Q \lor T) \lor R$ 11, *Assoc*
    13. ⎣ $Q \lor T$ 3, 12, $\lor E$
    14. $(P \lor S) \supset (Q \lor T)$ 5–13, *CP*

Exercise 3.13 (p. 145)

2. $P \equiv (Q \lor R)$ $\neg Q$ | $\neg P \supset R$

I: $\{P=F, Q=F, R=F\}$

4. 1.+ $(P \lor Q) \supset (R \equiv S)$
    2.+ $\neg(\neg S \land P)$
    3.+ $R \supset T$
    4. ? $P \supset (T \land R)$
    5.   ⎡ P
    6. | ? $T \land R$
    7. | $S \lor \neg P$ 2, *DeM*
    8. | S 5, 7, $\lor E$
    9. | $P \lor Q$ 5, $\lor I$
    10. | $R \equiv S$ 1, 9, *MP*
    11. | $(R \supset S) \land (S \supset R)$ 10, *Bicond*
    12. | $S \supset R$ 11, $\land E$
    13. | R 8, 12, *MP*
    14. | T 3, 13, *MP*
    15. ⎣ $T \land R$ 13, 14, $\land I$
    16. $P \supset (T \land R)$ 5–15, *CP*

6. 1.+ $\neg P \supset (Q \supset R)$
    2.+ $(P \vee S) \supset T$
    3.+ $R \supset (P \vee S)$
    4.+ $\neg T$
    5. ? $\neg Q$
    6. $\neg(P \vee S)$ 2, 4, MT
    7. $\neg P \wedge \neg S$ 6, DeM
    8. $\neg P$ 7, ∧E
    9. $Q \supset R$ 1, 8, MP
    10. $\neg R$ 3, 6, MT
    11. $\neg Q$ 9, 10, MT

8.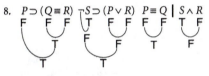

I: $\{P=F, Q=F, R=F, S=T\}$

10. 1.+ $P \supset (Q \supset R)$
     2.+ $\neg R$
     3. ? $P \supset \neg Q$
     4. ┌ P
     5. │ ? $\neg Q$
     6. │ $Q \supset R$ 1, 4, MP
     7. └ $\neg Q$ 2, 6, MT
     8. $P \supset \neg Q$ 4–7, CP

Exercise 3.14 (p. 149)

2. ⊢ $P \supset (\neg P \supset P)$
    1. ┌ P
    2. │ ? $\neg P \supset P$
    3. │ $P \vee P$ 1, Taut
    4. └ $\neg P \supset P$ 3, Cond
    5. $P \supset (\neg P \supset P)$ 1–4, CP

4. ⊢ $((P \supset Q) \wedge \neg Q) \supset \neg P$
    1. ┌ $(P \supset Q) \wedge \neg Q$
    2. │ ? $\neg P$
    3. │ $P \supset Q$ 1, ∧E
    4. │ $\neg Q$ 1, ∧E
    5. └ $\neg P$ 3, 4, MT
    6. $((P \supset Q) \wedge \neg Q) \supset \neg P$ 1–5, CP

6. ⊢ $(P \vee Q) \equiv \neg(\neg P \wedge \neg Q)$
    1. ┌ $P \vee Q$
    2. │ ? $\neg(\neg P \wedge \neg Q)$
    3. └ $\neg(\neg P \wedge \neg Q)$ 1, DeM
    4. $(P \vee Q) \supset \neg(\neg P \wedge \neg Q)$ 1–3, CP
    5. ┌ $\neg(\neg P \wedge \neg Q)$
    6. │ ? $P \vee Q$
    7. └ $P \vee Q$ 5, DeM
    8. $\neg(\neg P \wedge \neg Q) \supset (P \vee Q)$ 5–7, CP
    9. $((P \vee Q) \supset \neg(\neg P \wedge \neg Q)) \wedge$
          $(\neg(\neg P \wedge \neg Q) \supset (P \vee Q))$ 4, 8, ∧I
    10. $(P \vee Q) \equiv \neg(\neg P \wedge \neg Q)$ 9, Bicond

8. ⊢ $(P \supset Q) \vee (Q \supset P)$
    1. ┌ $\neg((P \supset Q) \vee (Q \supset P))$
    2. │ ? X
    3. │ $\neg((\neg P \vee Q) \vee (Q \supset P))$ 1, Cond
    4. │ $\neg(\neg P \vee Q) \wedge \neg(Q \supset P)$ 3, DeM
    5. │ $\neg(\neg P \vee Q)$ 4, ∧E
    6. │ $P \wedge \neg Q$ 5, DeM
    7. │ $\neg(Q \supset P)$ 4, ∧E
    8. │ $\neg(\neg Q \vee P)$ 7, Cond
    9. │ $Q \wedge \neg P$ 8, DeM
    10. │ P 6, ∧E
    11. │ $\neg P$ 9, ∧E
    12. └ $P \wedge \neg P$ 10, 11, ∧I
    13. $(P \supset Q) \vee (Q \supset P)$ 1–12, IP

10. $\vdash \neg(P \supset Q) \equiv (P \wedge \neg Q)$

    1. $\neg P \supset Q$

    2. $?\ P \wedge \neg Q$

    3. $\neg(\neg P \vee Q)$ 1, *Cond*

    4. $P \wedge \neg Q$ 3, *DeM*

    5. $\neg(P \supset Q) \supset (P \wedge \neg Q)$ 1–4, *CP*

    6. $P \wedge \neg Q$

    7. $?\ \neg(P \supset Q)$

    8. $\neg(\neg P \vee Q)$ 6, *DeM*

    9. $\neg(P \supset Q)$ 8, *Cond*

    10. $(P \wedge \neg Q) \supset \neg(P \supset Q)$ 6–9, *CP*

    11. $(\neg(P \supset Q) \supset (P \wedge \neg Q)) \wedge ((P \wedge \neg Q) \supset \neg(P \supset Q))$ 5, 10, $\wedge I$

    12. $\neg(P \supset Q) \equiv (P \wedge \neg Q)$ 11, *Bicond*

CHAPTER FOUR EXERCISES

Exercises 4.00 (p. 160)

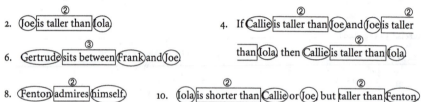

2. Joe is taller than Iola

6. Gertrude sits between Frank and Joe

8. Fenton admires himself.

4. If Callie is taller than Joe and Joe is taller than Iola, then Callie is taller than Iola

10. Iola is shorter than Callie or Joe, but taller than Fenton

Exercises 4.01 (p. 165)

2. Tji

6. $Bgfj$

10. $(Sic \wedge Sij) \wedge Tif$

4. $(Tcj \wedge Tji) \supset Tci$

8. Aff

Exercises 4.02 (p. 173)

2. $\neg Sg \wedge Ag$

6. $Cf \wedge Sf$

10. $(\forall x)(Ax \supset Cx) \supset Cg$

4. $(\exists x)(Ax \wedge Sx) \supset (\exists x)(Sx \wedge Ax)$

8. $(\forall x)(Ax \supset Cx)$

Exercises 4.03 (pp. 175–76)

2. $(\exists x)(Fx \wedge Gx)$

6. $(\exists y)((Fy \wedge Gy) \wedge (Hy \wedge Iy))$

10. $(\forall x)(Fx \supset Gx) \supset Ha$

4. $(\exists y)(Fx \wedge Gx) \wedge (\forall x)(Fy \supset Gy)$

8. $(((\exists y)Fy \wedge Gy) \wedge (Hy \wedge Iy))$

Exercises 4.04 (pp. 178–79)

2. $Kc \supset (\exists x)(Sx \wedge \neg Wx)$
6. $\neg(\exists x)(Sx \wedge \neg Wx)$
10. $\neg(Kg \vee Kc) \supset Wi$

4. $(\forall x)(Sx \supset Wx) \supset (\neg Wf \supset \neg Sf)$
8. $(\exists x)(Sx \wedge Kx) \supset \neg(\forall x)(Sx \supset Wx)$

Exercises 4.05 (p. 182)

2. $(\exists x)(Sx \wedge Cxf)$
6. $\neg Eg \supset \neg(\exists x)((Sx \wedge Cx) \wedge Ex)$
10. $(\exists x)((Sx \wedge Cx) \wedge Ex) \supset (\exists x)((Sx \wedge Cx) \wedge Cxg)$

4. $(\forall x)(Sx \supset Cxc)$
8. $\neg(\forall x)((Sx \wedge Cx) \supset Ex)$

Exercises 4.06 (p. 187)

2. $\neg(\exists x)(Sx \wedge (\exists y)(Cy \wedge Axy))$
6. $(\forall x)((Cx \wedge \neg Mx) \supset Afx)$
10. $\neg(\forall x)((Sx \wedge Mx) \supset (\exists y)((Cy \wedge \neg My) \wedge \neg Axy))$

4. $(\exists x)(Sx \wedge (\forall y)(Cy \supset Axy))$
8. $\neg(\forall x)(Sx \supset (\exists y)(Cy \wedge Axy))$

Exercises 4.07 (p. 193)

2. $(\exists x)(Px \wedge (\forall y)(Sy \supset Ayx))$
6. $\neg(Sf \wedge Wf) \supset \neg(\exists x)(Px \wedge Axf)$
10. $(\forall x)((Sx \wedge Wx) \supset \neg(\exists y)Axy)$

4. $\neg(\exists x)(Sx \wedge (\forall y)(Py \supset Ayx))$
8. $Acf \supset (\exists x)((Sx \wedge Wx) \wedge Acx)$

Exercises 4.08 (pp. 197–98)

2. $g = h$
6. $j \neq f$
10. $Ahj \vee h \neq g$

4. $(\exists x)((Axj \wedge (\forall y)(Ayj \supset x = y)) \wedge \neg Sx)$
8. $g = h \supset (Sg \supset Sh)$

Exercise 4.09 (p. 206)

2. $(\exists x)(Sx \wedge Wx)$
4. $(\exists x)(((Sx \wedge Wx) \wedge (\forall y)((Sy \wedge Wy) \supset x = y)) \wedge (\exists z)(Pz \wedge Axz))$
6. $(\exists x)(((Sx \wedge Wx) \wedge (\forall y)((Sy \wedge Wy) \supset x = y)) \wedge Agx)$
8. $(\forall x)(\forall y)((((Sx \wedge Wx) \wedge (Sy \wedge Wy)) \wedge x \neq y) \supset (\forall z)((Sz \wedge Wz) \supset (x = z \vee y = z))$
10. $(\exists x)(\exists y)((((Sx \wedge Sy) \wedge x \neq y) \wedge (\forall z)(Sz \supset (x = z \vee y = z))) \wedge (Afx \wedge Afy))$

Exercise 4.10 (p. 210)

2. $(\exists x)(Sx \wedge \neg Tfx)$
6. $(Sc \wedge Sf) \wedge (\forall x)((Sx \wedge (x \neq c \wedge x \neq f)) \supset Tfx)$
10. $Sf \supset \neg(\forall x)((Sx \wedge x \neq f) \supset Tfx)$

4. $\neg(\exists x)((Sx \wedge x \neq c) \wedge Txc)$
8. $(\exists x)((Sx \wedge (\forall y)((Sy \wedge s \neq y) \supset Txy)) \wedge Txc)$

Exercises 4.11 (pp. 214)

2. $(\forall x)(Sx \supset (\exists y)((Sy \wedge By) \wedge Axy))$ $\Rightarrow$ $(\forall x)(\exists y)(By \wedge Axy)$
4. $(\exists x)((Sx \wedge Bx) \wedge (\forall y)(Sy \wedge Axy))$ $\Rightarrow$ $(\exists x)(Bx \wedge (\forall y)Axy)$
6. $(\forall x)((Sx \wedge Bx) \supset (\forall y)((Sy \wedge By) \supset x = y))$ $\Rightarrow$ $(\forall x)(Bx \supset (\forall y)(By \supset x = y))$
8. $(\forall x)(\forall y)(((Sx \wedge Sy) \wedge x \neq y) \supset (\forall z)(Sz \supset (x = z \vee y = z)))$ $\Rightarrow$

                                                   $(\forall x)(\forall y)(x \neq y \supset (\forall z)(x = z \vee y = z))$

10. $\neg(\exists x)(Sx \wedge (\exists y)((Sy \wedge \neg By) \wedge Axy))$ $\Rightarrow$ $\neg(\exists x)(\exists y)(\neg By \wedge Axy)$

CHAPTER FIVE EXERCISESS

Exercises 5.00 (p. 231)

2.
1.+	$(\exists x)Fx \wedge (\exists x)Gx$	
2.+	$(\exists x)Hx \supset \neg(\exists x)Gx$	
3. ?	$\neg(\exists x)Hx$	
4.	$(\exists x)Gx$	1, $\wedge E$
5.	$\neg(\exists x)Hx$	2, 4, MT

4.
1.+	$(\exists x)Fx \supset (\exists x)Gx$	
2.+	$\neg(\exists x)Gx \vee (\forall x)Fx$	
3. ?	$(\exists x)Fx \supset (\forall x)Fx$	
4.	$(\exists x)Gx \supset (\forall x)Fx$	2, Cond
5.	$(\exists x)Fx \supset (\forall x)Fx$	1, 4, HS

6.
1.+	$(\forall x)(Fx \supset Gx)$	
2. ?	$(\forall x)\neg(Fx \wedge \neg Gx)$	
3.	$(\forall x)(\neg Fx \vee Gx)$	1, Cond
4.	$(\forall x)\neg(Fx \wedge \neg Gx)$	3, DeM

8.
1.+	$(\exists x)((Fx \wedge \neg Gx) \vee \neg Gx)$	
2. ?	$(\exists x)((Fx \vee \neg Gx) \wedge \neg Gx)$	
3.	$(\exists x)(\neg Gx \vee (Fx \wedge \neg Gx))$	1, Com
4.	$(\exists x)((\neg Gx \vee Fx) \wedge (\neg Gx \vee \neg Gx))$	3, Dist
5.	$(\exists x)((\neg Gx \vee Fx) \wedge \neg Gx)$	4, Taut
6.	$(\exists x)((Fx \vee \neg Gx) \wedge \neg Gx)$	5, Com

10.
1.+	$(\forall x)Fx \supset \neg(\exists y)Gy$	
2.+	$\neg(\exists x)Hx \supset (\exists y)Gy$	
3. ?	$(\forall x)Fx \supset (\exists x)Hx$	
4.	$\neg(\exists y)Gy \supset (\exists x)Hx$	2, Contra
5.	$(\forall x)Fx \supset (\exists x)Hx$	1, 4, HS

Exercises 5.01 (p. 236)

2.
1.+	$(\forall x)(Fx \supset Gx)$	
2. ?	$\neg(\exists x)(Fx \wedge \neg Gx)$	
3.	$\neg(\exists x)\neg(Fx \supset Gx)$	1, QT
4.	$\neg(\exists x)\neg(\neg Fx \vee Gx)$	3, Cond
5.	$\neg(\exists x)(Fx \wedge \neg Gx)$	4, DeM

4.
1.+	$(\exists x)((Fx \wedge Gx) \vee \neg Hx)$	
2. ?	$\neg(\forall x)(Hx \wedge (Fx \supset \neg Gx))$	
3.	$\neg(\forall x)\neg((Fx \wedge Gx) \vee \neg Hx)$	1, QT
4.	$\neg(\forall x)(\neg(Fx \wedge Gx) \wedge Hx)$	3, DeM
5.	$\neg(\forall x)(Hx \wedge \neg(Fx \wedge Gx))$	4, Com
6.	$\neg(\forall x)(Hx \wedge (\neg Fx \vee \neg Gx))$	5, DeM
7.	$\neg(\forall x)(Hx \wedge (Fx \supset \neg Gx))$	6, Cond

6.
1.+	$(\forall x)((Fx \supset Gx) \wedge Hx)$	
2. ?	$\neg(\exists x)((\neg Hx \vee Fx) \wedge (\neg Hx \vee \neg Gx))$	
3.	$\neg(\exists x)\neg((Fx \supset Gx) \wedge Hx)$	1, QT
4.	$\neg(\exists x)(\neg(Fx \supset Gx) \vee \neg Hx)$	3, DeM
5.	$\neg(\exists x)(\neg(\neg Fx \vee Gx) \vee \neg Hx)$	4, Cond
6.	$\neg(\exists x)((Fx \wedge \neg Gx) \vee \neg Hx)$	5, DeM
7.	$\neg(\exists x)(\neg Hx \vee (Fx \wedge \neg Gx))$	6, Com
8.	$\neg(\exists x)((\neg Hx \vee Fx) \wedge (\neg Hx \vee \neg Gx))$	7, Dist

8.
1.+	$(\forall x)Fx$	
2. ?	$\neg((\forall x)Fx \supset \neg(\forall x)Fx)$	
3.	$(\forall x)Fx \wedge (\forall x)Fx$	1, Taut
4.	$\neg(\neg(\forall x)Fx \vee \neg(\forall x)Fx)$	3, DeM
5.	$\neg((\forall x)Fx \supset \neg(\forall x)Fx)$	4, Cond

10.
1.+	$(\exists x)\neg Fx$	
2. ?	$\neg(\neg(\forall x)Fx \supset (\forall x)Fx)$	
3.	$\neg(\forall x)Fx$	1, QT
4.	$\neg(\forall x)Fx \wedge \neg(\forall x)Fx$	3, Taut
5.	$\neg((\forall x)Fx \vee (\forall x)Fx)$	4, DeM
6.	$\neg(\neg(\forall x)Fx \supset (\forall x)Fx)$	5, Cond

Exercises 5.02 (pp. 240–41)

2.	1.+	$(\forall x)(Fx \supset Gx)$	
	2.+	$(\forall x)(Gx \supset Hx)$	
	3. ?	$Fa \supset Ha$	
	4.	$Fa \supset Ga$	1, UI
	5.	$Ga \supset Ha$	2, UI
	6.	$Fa \supset Ha$	4, 5, HS

4.	1.+	$(\forall x)(Fx \supset Gx)$	
	2.+	$\neg(\exists x)(Gx \wedge \neg Hx)$	
	3. ?	$\neg(Fa \wedge \neg Ha)$	
	4.	$Fa \supset Ga$	1, UI
	5.	$(\forall x)\neg(Gx \wedge \neg Hx)$	2, QT
	6.	$\neg(Ga \wedge \neg Ha)$	5, UI
	7.	$\neg Ga \vee Ha$	6, DeM
	8.	$Ga \supset Ha$	7, Cond
	9.	$Fa \supset Ha$	4, 8, HS
	10.	$\neg Fa \vee Ha$	9, Cond
	11.	$\neg(Fa \wedge \neg Ha)$	10, DeM

6.	1.+	$(\forall x)Fx \supset (\forall x)Gx$	
	2.+	$(\forall x)\neg Gx$	
	3. ?	$(\exists x)\neg Fx$	
	4.	$\ulcorner \neg(\exists x)\neg Fx$	
	5.	$? X$	
	6.	$(\forall x)Fx$	4, QT
	7.	$(\forall x)Gx$	1, 6, MP
	8.	Ga	7, UI
	9.	$\neg Ga$	2, UI
	10.	$\llcorner Ga \wedge \neg Ga$	8, 9, $\wedge$I
	11.	$(\exists x)\neg Fx$	4–10, IP

8.	1.+	$(\forall x)Fx$	
	2.+	$(\forall x)(Fx \supset Gx)$	
	3. ?	$(\exists x)(Fx \wedge Gx)$	
	4.	$\ulcorner \neg(\exists x)(Fx \wedge Gx)$	
	5.	$? X$	
	6.	$(\forall x)\neg(Fx \wedge Gx)$	4, QT
	7.	$\neg(Fa \wedge Ga)$	6, UI
	8.	$\neg Fa \vee \neg Ga$	7, DeM
	9.	Fa	1, UI
	10.	$\neg Ga$	8, 9, $\vee$E
	11.	$Fa \supset Ga$	2, UI
	12.	$\neg Fa$	10, 11, MT
	13.	$\llcorner Fa \wedge \neg Fa$	9, 12, $\wedge$I
	14.	$(\exists x)(Fx \wedge Gx)$	4–13, IP

10.	1.+	$\neg(\exists x)(\neg Fx \vee Hx)$	
	2.+	$(\forall x)(Jx \supset Gx)$	
	3.+	$(\forall x)(Fx \supset Jx)$	
	4. ?	$(\exists x)(Fx \wedge Gx)$	
	5.	$\ulcorner \neg(\exists x)(Fx \wedge Gx)$	
	6.	$? X$	
	7.	$(\forall x)\neg(Fx \wedge Gx)$	5, QT
	8.	$\neg(Fa \wedge Ga)$	7, UI
	9.	$\neg Fa \vee \neg Ga$	8, DeM
	10.	$(\forall x)\neg(\neg Fx \vee Hx)$	1, QT
	11.	$\neg(\neg Fa \vee Ha)$	10, UI
	12.	$Fa \wedge \neg Ha$	11, DeM
	13.	$Fa \supset Ja$	3, UI
	14.	$Ja \supset Ga$	2, UI
	15.	$Fa \supset Ga$	13, 14, HS
	16.	Fa	12, $\wedge$E
	17.	Ga	15, 16, MP
	18.	$\neg Ga$	9, 16, $\vee$E
	19.	$\llcorner Ga \wedge \neg Ga$	17, 18, $\wedge$I
	20.	$(\exists x)(Fx \wedge Gx)$	5–19, IP

Exercises 5.03 (pp. 245–46)

2. 1.+ Fa
    2.+ $(\exists x)Fx \supset (\forall x)(Gx \vee Hx)$
    3.+ $(\exists x)(Gx \vee Hx) \supset Ha$
    4. ? $(\exists x)Hx$
    5. $(\exists x)Fx$ 1, EG
    6. $(\forall x)(Gx \vee Hx)$ 2, 5, MP
    7. $Ga \vee Ha$ 6, UI
    8. $(\exists x)(Gx \vee Hx)$ 7, EG
    9. Ha 3, 8, MP
    10. $(\exists x)Hx$ 9, EG

4. 1.+ $(\exists x)Fx \supset (\forall x)(Gx \supset Hx)$
    2.+ $(\forall x)(Fx \supset Gx)$
    3. ? $Fa \supset (Ga \wedge Ha)$
    4. ⌈ Fa
    5. | ? $Ga \wedge Ha$
    6. | $(\exists x)Fx$ 4, EG
    7. | $(\forall x)(Gx \supset Hx)$ 1, 6, MP
    8. | $Ga \supset Ha$ 7, UI
    9. | $Fa \supset Ga$ 2, UI
    10. | $Fa \supset Ha$ 8, 9, HS
    11. | Ga 4, 9, MP
    12. | Ha 4, 10, MP
    13. ⌊ $Ga \wedge Ha$ 11, 12, $\wedge I$
    14. $Fa \supset (Ga \wedge Ha)$ 4–13, CP

6. 1.+ $(\forall x)(\neg Gx \supset Hx)$
    2.+ $(\forall x)\neg(Fx \supset Hx)$
    3. ? $(\exists x)(Fx \wedge Gx)$
    4. $\neg Ga \supset Ha$ 1, UI
    5. $\neg(Fa \supset Ha)$ 2, UI
    6. $\neg(\neg Fa \vee Ha)$ 5, Cond
    7. $Fa \wedge \neg Ha$ 6, DeM
    8. Fa 7, $\wedge E$
    9. $\neg Ha$ 7, $\wedge E$
    10. Ga 4, 9, MT
    11. $Fa \wedge Ga$ 8, 10, $\wedge I$
    12. $(\exists x)(Fx \wedge Gx)$ 11, EG

8. 1.+ $(\forall x)(Fx \wedge Hx)$
    2.+ $(\exists x)(Gx \vee Ix) \supset Ja$
    3.+ $Ja \supset (\forall x)(Gx \supset \neg Fx)$
    4. ? $\neg(\forall x)(Fx \supset Gx)$
    5. ⌈ $(\forall x)(Fx \supset Gx)$
    6. | ? X
    7. | $Fa \supset Ga$ 5, UI
    8. | $Fa \wedge Ha$ 1, UI
    9. | Fa 8, $\wedge E$
    10. | Ga 7, 9, MP
    11. | $Ga \vee Ia$ 10, $\vee I$
    12. | $(\exists x)(Gx \vee Ix)$ 11, EG
    13. | Ja 2, 12, MP
    14. | $(\forall x)(Gx \supset \neg Fx)$ 3, 13, MP
    15. | $Ga \supset \neg Fa$ 14, UI
    16. | $\neg Fa$ 10, 15, MP
    17. ⌊ $Fa \wedge \neg Fa$ 9, 16, $\wedge I$
    18. $\neg(\forall x)(Fx \supset Gx)$ 5–17, IP

10. 1.+ $(\forall x)(Gx \supset \neg Hx)$
    2.+ $(\forall x)(Fx \vee Gx)$
    3. ? $(\forall x)Hx \supset (\exists x)Fx$
    4. ⌈ $(\forall x)Hx$
    5. | ? $(\exists x)Fx$
    6. | $Ga \supset \neg Ha$ 1, UI
    7. | Ha 4, UI
    8. | $\neg Ga$ 6, 7, MT
    9. | $Fa \vee Ga$ 2, UI
    10. | Fa 8, 9, $\vee E$
    11. ⌊ $(\exists x)Fx$ 10, EG
    12. $(\forall x)Hx \supset (\exists x)Fx$ 4–11, CP

Exercises 5.04 (pp. 250–51)

2. 1.+ $(\exists x)(\exists y)Fxy$
    2.+ $(\forall x)(\forall y)(Fxy \supset Gx)$
    3. ? $(\exists x)Gx$
    4. $(\exists y)Fay$ 1, EI
    5. Fab 4, EI
    6. $(\forall y)(Fay \supset Ga)$ 2, UI
    7. $Fab \supset Ga$ 6, UI
    8. Ga 5, 7, MP

4. 1.+ $\neg(\forall x)Gx \supset (\forall x)Hx$
    2.+ $(\exists x)Hx \supset (\forall x)\neg Fx$
    3. ? $(\forall x)(Fx \supset Gx)$
    4. ⌈ $\neg(\forall x)(Fx \supset Gx)$
    5. │ ? X
    6. │ $(\exists x)\neg(Fx \supset Gx)$ 4, QT
    7. │ $\neg(Fa \supset Ga)$ 6, EI
    8. │ $\neg(\neg Fa \vee Ga)$ 7, Cond
    9. │ $Fa \wedge \neg Ga$ 8, DeM
    10. │ Fa 9, ∧E
    11. │ $(\exists x)Fx$ 10, EG
    12. │ $\neg(\forall x)\neg Fx$ 11, QT
    13. │ $\neg(\exists x)Hx$ 2, 12, MT
    14. │ $\neg Ga$ 9, ∧E
    15. │ $(\exists x)\neg Gx$ 14, EG
    16. │ $\neg(\forall x)Gx$ 15, QT
    17. │ $(\forall x)Hx$ 1, 16, MP
    18. │ Ha 17, UI
    19. │ $(\forall x)\neg Hx$ 13, QT
    20. │ $\neg Ha$ 19, UI
    21. ⌊ $Ha \wedge \neg Ha$ 18, 20, ∧I
    22. $(\forall x)(Fx \supset Gx)$ 4–21, IP

6. 1.+ $(\forall x)(Fx \supset (\exists y)Gy)$
    2.+ $(\exists y)Gy \supset Ha$
    3. ? $(\exists x)Fx \supset (\exists x)Hx$
    4. ⌈ $(\exists x)Fx$
    5. │ ? $(\exists x)Hx$
    6. │ Fb 4, EI
    7. │ $Fb \supset (\exists y)Gy$ 1, UI
    8   │ $(\exists y)Gy$ 6, 7, MP
    9. │ Ha 2, 8, MP
    10. ⌊ $(\exists x)Hx$ 9, EG
    11. $(\exists x)Fx \supset (\exists x)Hx$ 4–10, CP

8. 1.+ $(\exists x)(Fx \vee Gx)$
    2.+ $(\forall x)(Fx \supset Hx)$
    3.+ $(\forall x)(Gx \supset Hx)$
    4. ? $(\exists x)Hx$
    5. $Fa \vee Ga$ 1, EI
    6. $Fa \supset Ha$ 2, UI
    7. $Ga \supset Ha$ 3, UI
    8. $Ha \vee Ha$ 5, 6, 7, CD
    9. Ha 8, Taut
    10. $(\exists x)Hx$ 9, EG

10. 1.+ $(\exists x)(\neg Fx \vee Hx)$
    2.+ $(\forall x)(Fx \supset (Gx \supset Hx))$
    3. ? $(\exists x)(Fx \wedge Gx) \supset (\exists x)Hx$
    4. ⌈ $(\exists x)(Fx \wedge Gx)$
    5. │ ? $(\exists x)Hx$
    6. │ $Fa \wedge Ga$ 4, EI
    7. │ $Fa \supset (Ga \supset Ha)$ 2, UI
    8. │ $(Fa \wedge Ga) \supset Ha$ 7, Exp
    9. │ Ha 6, 8, MP
    10. ⌊ $(\exists x)Hx$ 9, EG
    11. $(\exists x)(Fx \wedge Gx) \supset (\exists x)Hx$ 4–10, CP

Exercises 5.05 (pp. 258–59)

2. 1.+ $(\forall x)(Fx \supset Gx)$
    2.+ $(\exists x)Gx \supset (\forall x)(Fx \supset Hx)$
    3. ? $(\forall x)Fx \supset (\forall x)Hx$
    4. ⌈ $(\forall x)Fx$
    5. │ ? $(\forall x)Hx$
    6. │ Fa 4, UI
    7. │ $Fa \supset Ga$ 1, UI
    8. │ Ga 6, 7, MP
    9. │ $(\exists x)Gx$ 8, EG
    10. │ $(\forall x)(Fx \supset Hx)$ 2, 9, MP
    11. │ $Fa \supset Ha$ 10, UI
    12. │ Ha 6, 11, MP
    13. ⌊ $(\forall x)Hx$ 12, UG
    14. $(\forall x)Fx \supset (\forall x)Hx$ 4–13, CP

4. 1.+ $(\forall x)(Fx \wedge Gx)$
    2. ? $(\forall x)Fx \wedge (\forall x)Gx$
    3. $Fa \wedge Ga$ 1, UI
    4. Fa 3, ∧E
    5. Ga 3, ∧E
    6. $(\forall x)Fx$ 4, UG
    7. $(\forall x)Gx$ 5, UG
    8. $(\forall x)Fx \wedge (\forall x)Gx$ 6, 7, ∧I

6. 1.+ $(\exists x)(Fx \lor Hx)$
    2.+ $(\forall x)(Hx \supset Fx)$
    3.+ $(\exists x)Gx \supset (\forall x)(Gx \supset Hx)$
    4. ? $(\forall x)(Gx \supset Fx)$
    5. ⎡ Ga
    6. │ $?Fa$
    7. │ $(\exists x)Gx$ 5, EG
    8. │ $(\forall x)(Gx \supset Hx)$ 3, 7, MP
    9. │ $Ga \supset Ha$ 8, UI
    10. │ Ha 5, 9, MP
    11. │ $Ha \supset Fa$ 2, UI
    12. ⎣ Fa 10, 11, MP
    13. $Ga \supset Fa$ 5–12, CP
    14. $(\forall x)(Gx \supset Fx)$ 13, UG

8. 1.+ $(\forall x)(Fx \supset Gx)$
    2.+ $(\forall x)(\exists y)(Fy \land Hxy)$
    3. ? $(\forall x)(\exists y)(Gy \land Hxy)$
    4. $(\exists y)(Fy \land Hay)$ 2, UI
    5. $Fb \land Hab$ 4, EI
    6. $Fb \supset Gb$ 1, UI
    7. Fb 5, $\land$E
    8. Gb 6, 7, MP
    9. Hab 5, $\land$E
    10. $Gb \land Hab$ 8, 9, $\land$I
    11. $(\exists y)(Gy \land Hay)$ 10, EG
    12. $(\forall x)(\exists y)(Gy \land Hxy)$ 11, UG

10. 1.+ $(\forall x)(\forall y)(Fxy \supset (\exists z)Gzxy)$
    2.+ $(\forall x)(\forall y)(\forall z)(Gzxy \supset (Hzx \land Hzy))$
    3.+ $(\forall x)Fxa$
    4. ? $(\forall x)(\exists y)(Gyxa \supset Hya)$
    5. $(\forall y)(Fby \supset (\exists z)Gzby)$ 1, UI
    6. $Fba \supset (\exists z)Gzba$ 5, UI
    7. Fba 3, UI
    8. $(\exists z)Gzba$ 6, 7, MP
    9. $Gcba$ 8, EI
    10. ⎡ $Gcba$
    11. │ $? Hca$
    12. │ $(\forall y)(\forall z)(Gzby \supset (Hzb \land Hzy))$ 2, UI
    13. │ $(\forall z)(Gzba \supset (Hzb \land Hza))$ 12, UI
    14. │ $Gcba \supset (Hcb \land Hca)$ 13, UI
    15. │ $Hcb \land Hca$ 10, 14, MP
    16. ⎣ Hca 15, $\land$E
    17. $Gcba \supset Hca$ 10–16, CP
    18. $(\exists y)(Gyba \supset Hya)$ 17, EG
    19. $(\forall x)(\exists y)(Gyxa \supset Hya)$ 18, UG

Exercises 5.06 (pp. 263–64)

2. 1.+ $(\exists x)(Fx \land Gx)$
    2.+ $(\exists x)(Fx \land \neg Gx)$
    3. ? $(\exists x)(\exists y)((Fx \land Fy) \land x \neq y)$
    4. $Fa \land Ga$ 1, EI
    5. $Fb \land \neg Gb$ 2, EI
    6. ⎡ $a = b$
    7. │ $? X$
    8. │ Ga 4, $\land$E
    9. │ $\neg Gb$ 5, $\land$E
    10. │ $\neg Ga$ 6, 9, ID
    11. ⎣ $Ga \land \neg Ga$ 8, 10, $\land$I
    12. $a \neq b$ 6–11, IP
    13. Fa 4, $\land$E
    14. Fb 5, $\land$E
    15. $Fa \land Fb$ 13, 14, $\land$I

4. 1.+ $(\forall x)(Fx \supset (\forall y)(Fy \supset x = y))$
    2.+ $(\exists x)(Fx \land Gx)$
    3. ? $(\forall x)(Fx \supset Gx)$
    4. ⎡ Fa
    5. │ $? Ga$
    6. │ $Fb \land Gb$ 2, EI
    7. │ $Fa \supset (\forall y)(Fy \supset a = y)$ 1, UI
    8. │ $(\forall y)(Fy \supset a = y)$ 4, 7, MP
    9. │ $Fb \supset a = b$ 8, UI
    10. │ Fb 6, $\land$E
    11. │ $a = b$ 9, 10, MP
    12. │ Gb 6, $\land$E
    13. ⎣ Ga 11, 12, ID
    14. $Fa \supset Ga$ 4–13, CP
    15. $(\forall x)(Fx \supset Gx)$ 14, UG

16. $(Fa \wedge Fb) \wedge a \neq b$ 12, 15, $\wedge I$
17. $(\exists y)((Fa \wedge Fy) \wedge a \neq y)$ 16, EG
18. $(\exists x)(\exists y)((Fx \wedge Fy) \wedge x \neq y)$ 17, EG

6. 1.+ $(\exists x)(Fx \wedge (\forall y)(Fy \supset x = y))$
    2.+ $\neg Fb$
    3. ? $(\exists x)x \neq b$
    4. $\neg(\exists x)x \neq b$
    5.   ? X
    6. $Fa \wedge (\forall y)(Fy \supset a = y)$ 1, EI
    7. $(\forall y)(Fy \supset a = y)$ 6, $\wedge E$
    8. $Fb \supset a = b$ 7, UI
    9. $(\forall x)x = b$ 4, QT
    10. $a = b$ 9, UI
    11. Fa 6, $\wedge E$
    12. Fb 10, 11, ID
    13. $Fb \wedge \neg Fb$ 2, 12, $\wedge I$
    14. $(\exists x)x \neq b$ 4–13, IP

8. 1.+ $(\exists x)((Fx \wedge Gax) \wedge Hx)$
    2.+ $Fb \wedge Gab$
    3.+ $(\forall x)((Fx \wedge Gax) \supset x = b)$
    4. ? Hb
    5. $(Fc \wedge Gac) \wedge Hc$ 1, EI
    6. $(Fc \wedge Gac) \supset c = b$ 3, UI
    7. $Fc \wedge Gac$ 5, $\wedge E$
    8. $c = b$ 6, 7, MP
    9. Hc 5, $\wedge E$
    10. Hb 8, 9, ID

10. 1.+ $(\forall x)(\forall y)((Fxy \wedge x \neq y) \supset Gxy)$
     2.+ $(\exists x)(\forall y)(x \neq y \supset Fxy)$
     3. ? $(\exists x)(\forall y)(x \neq y \supset Gxy)$
     4. $(\forall y)(a \neq y \supset Fay$ 2, EI
     5. $a \neq b$
     6.   ? Gab
     7. $a \neq b \supset Fab$ 4, UI
     8. Fab 5, 7, MP
     9. $(\forall y)((Fay \wedge a \neq y) \supset Gay)$ 1, UI
     10. $(Fab \wedge a \neq b) \supset Gab$ 9, UI
     11. $Fab \wedge a \neq b$ 5, 8, $\wedge I$
     12. Gab 10, 11, MP
     13. $a \neq b \supset Gab$ 5–12, CP
     14. $(\forall y)(a \neq y \supset Gay)$ 13, UG
     15. $(\exists x)(\forall y)(x \neq y \supset Gxy)$ 14, EG

Exercises 5.07 (pp. 266–67)

2. $\vdash (\forall x)(Fx \vee \neg Fx)$
    1. $\neg(\forall x)(Fx \vee \neg Fx)$
    2.   ? X
    3. $(\exists x)\neg(Fx \vee \neg Fx)$ 1, QT
    4. $\neg(Fa \vee \neg Fa)$ 3, EI
    5. $Fa \wedge \neg Fa$ 4, DeM
    6. $(\forall x)(Fx \vee \neg Fx)$ 1–5, IP

4. $\vdash (\forall x)(\forall y)((Fx \wedge x = y) \supset Fy)$
    1. $Fa \wedge a = b$
    2.   ? Fb
    3. Fa 1, $\wedge E$
    4. $a = b$ 1, $\wedge E$
    5. Fb 3, 4, ID
    6. $(Fa \wedge a = b) \supset Fb$ 1–5, CP
    7. $(\forall y)((Fa \wedge a = y) \supset Fy)$ 6, UG
    8. $(\forall x)(\forall y)((Fx \wedge x = y) \supset Fy)$ 7, UG

6. ⊢ $(\forall x)(\forall y)(x = y \supset (Fx \equiv Fy))$

  1. │ $a = b$
  2. │ ? $Fa \equiv Fb$
  3. │ ┌ Fa
  4. │ │ ? Fb
  5. │ └ Fb 1, 3, ID
  6. │ $Fa \supset Fb$ 3–5, CP
  7. │ ┌ Fb
  8. │ │ ? Fa
  9. │ └ Fa 1, 7, ID
  10. │ $Fb \supset Fa$ 7–9, CP
  11. │ $(Fa \supset Fb) \wedge (Fb \supset Fa)$ 6, 10, $\wedge I$
  12. │ $Fa \equiv Fb$ 11, Bicond
  13. $a = b \supset (Fa \equiv Fb)$ 1–12, CP
  14. $(\forall y)(a = y \supset (Fa \equiv Fy))$ 13, UG
  15. $(\forall x)(\forall y)(x = y \supset (Fx \equiv Fy))$ 14, UG

8. ⊢ $(\exists x)(\forall y)Fxy \supset (\forall y)(\exists x)Fxy$

  1. │ $(\exists x)(\forall y)Fxy$
  2. │ ? $(\forall y)(\exists x)Fxy$
  3. │ $(\forall y)Fay$ 1, EI
  4. │ Fab 3, UI
  5. │ $(\exists x)Fxb$ 4, EG
  6. │ $(\forall y)(\exists x)Fxy$ 5, UG
  $(\exists x)(\forall y)Fxy \supset (\forall y)(\exists x)Fxy$ 1–6, CP

10. ⊢ $(\forall x)(Fx \supset Gx \supset (\neg(\exists x)Gx \supset \neg(\exists x)Fx)$

  1. │ ┌ $(\forall x)(Fx \supset Gx)$
  2. │ │ ? $\neg(\exists x)Gx \supset \neg(\exists x)Fx$
  3. │ │ ┌ $\neg(\exists x)Gx$
  4. │ │ │ ? $\neg(\exists x)Fx$
  5. │ │ │ $Fa \supset Ga$ 1, UI
  6. │ │ │ $(\forall x)\neg Gx$ 3, QT
  7. │ │ │ $\neg Ga$ 6, UI
  8. │ │ │ $\neg Fa$ 5, 7, MT
  9. │ │ │ $(\forall x)\neg Fx$ 8, UG
  10. │ │ └ $\neg(\exists x)Fx$ 9, QT
  11. │ └ $\neg(\exists x)Gx \supset \neg(\exists x)Fx$ 3–10, CP
  12. $(\forall x)(Fx \supset Gx \supset (\neg(\exists x)Gx \supset \neg(\exists x)Fx)$ 1–11, CP

Exercises 5.08 (p. 278)

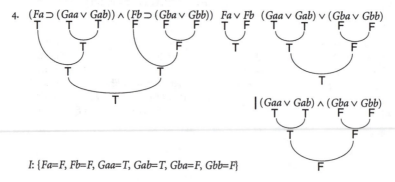

2. $Fa \supset Ga \quad \neg Fa \supset Ha \mid \neg Ga \supset \neg Ha$

I: $\{Fa=F, Ga=F, Ha=T\}$

4. $(Fa \supset (Gaa \vee Gab)) \wedge (Fb \supset (Gba \vee Gbb)) \quad Fa \vee Fb \quad (Gaa \vee Gab) \vee (Gba \vee Gbb)$

  $\mid (Gaa \vee Gab) \wedge (Gba \vee Gbb)$

I: $\{Fa=F, Fb=F, Gaa=T, Gab=T, Gba=F, Gbb=F\}$

6. $(Fa \supset (Gaa \lor Gba)) \lor (Fb \supset (Gab \lor Gbb))$ $(Fa \land (Haa \lor Hab)) \land (Fb \land (Hba \lor Hbb))$

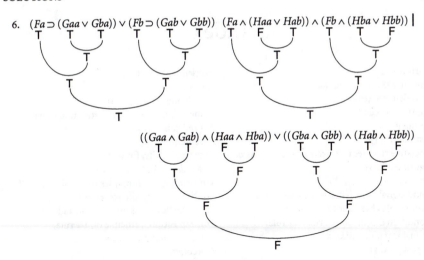

$((Gaa \land Gab) \land (Haa \land Hba)) \lor ((Gba \land Gbb) \land (Hab \land Hbb))$

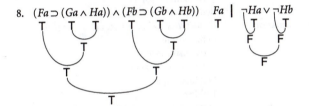

I: {Fa=T, Fb=T, Gaa=T, Gab=T, Gba=T, Gbb=T, Haa=F, Hab=T, Hba=T, Hbb=F}

8. $(Fa \supset (Ga \land Ha)) \land (Fb \supset (Gb \land Hb))$ Fa | $\neg Ha \lor \neg Hb$

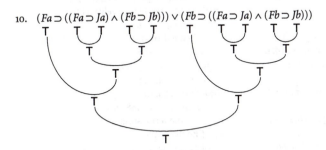

I: {Fa=T, Fb=T, Ga=T, Gb=T, Ha=T, Hb=T}

10. $(Fa \supset ((Fa \supset Ja) \land (Fb \supset Jb))) \lor (Fb \supset ((Fa \supset Ja) \land (Fb \supset Jb)))$

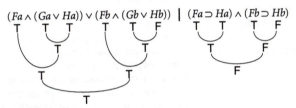

$(Fa \land (Ga \lor Ha)) \lor (Fb \land (Gb \lor Hb))$ | $(Fa \supset Ha) \land (Fb \supset Hb)$

I: {Fa=T, Fb=T, Ga=T, Gb=T, Ha=T, Hb=F, Ja=T, Jb=T}

Index